T0275727

Friction Stir Welding of 2XXX Aluminum Alloys Including Al–Li Alloys

Friction Stir Welding of 2XXX Aluminum Alloys Including Al–Li Alloys

A Volume in the Friction Stir Welding and Processing Book Series

Rajiv S. Mishra and Harpreet Sidhar
Department of Materials Science and Engineering,
University of North Texas, Denton, TX, USA

AMSTERDAM • BOSTON • HEIDELBERG • LONDON
NEW YORK • OXFORD • PARIS • SAN DIEGO
SAN FRANCISCO • SINGAPORE • SYDNEY • TOKYO

Butterworth-Heinemann is an imprint of Elsevier

Butterworth-Heinemann is an imprint of Elsevier
The Boulevard, Langford Lane, Kidlington, Oxford OX5 1GB, United Kingdom
50 Hampshire Street, 5th Floor, Cambridge, MA 02139, United States

Notices
Knowledge and best practice in this field are constantly changing. As new research and
experience broaden our understanding, changes in research methods, professional practices,
or medical treatment may become necessary.

Practitioners and researchers must always rely on their own experience and knowledge in
evaluating and using any information, methods, compounds, or experiments described herein.
In using such information or methods they should be mindful of their own safety and the safety
of others, including parties for whom they have a professional responsibility.

To the fullest extent of the law, neither the Publisher nor the authors, contributors, or editors,
assume any liability for any injury and/or damage to persons or property as a matter of products
liability, negligence or otherwise, or from any use or operation of any methods, products,
instructions, or ideas contained in the material herein.

British Library Cataloguing-in-Publication Data
A catalogue record for this book is available from the British Library

Library of Congress Cataloging-in-Publication Data
A catalog record for this book is available from the Library of Congress

ISBN: 978-0-12-805368-3

For Information on all Butterworth-Heinemann publications
visit our website at https://www.elsevier.com

Working together
to grow libraries in
developing countries

www.elsevier.com • www.bookaid.org

Publisher: Joe Hayton
Acquisition Editor: Christina Gifford
Senior Editorial Project Manager: Kattie Washington
Production Project Manager: Kiruthika Govindaraju
Cover Designer: MPS

Typeset by MPS Limited, Chennai, India

CONTENTS

Preface to This Volume of Friction Stir Welding and Processing Book Series

This is the seventh volume in the recently launched short book series on friction stir welding and processing. As highlighted in the preface of the first book, the intention of this book series is to serve engineers and researchers engaged in advanced and innovative manufacturing techniques. Friction stir welding was invented more than 20 years back as a solid-state joining technique. In this period, friction stir welding has found a wide range of applications in joining of aluminum alloys. Although the fundamentals have not kept pace in all aspects, there is a tremendous wealth of information in the large volume of papers published in journals and proceedings. Recent publications of several books and review articles have furthered the dissemination of information.

This book is focused on friction stir welding of 2XXX alloys, a topic of great interest for practitioners of this technology in aerospace sector. 2XXX series alloys are among highest specific strength aluminum alloys and achieving high joint efficiency is a desired goal. The basic research towards this has reached a level that warrants compilation of the scientific knowledge. This volume provides a good summary of the current understanding and provides brief guideline for future research directions. It is intended to serve as a resource for both researchers and engineers dealing with the development of high efficiency structures. As stated in the previous volumes, this short book series on friction stir welding and processing will include books that advance both the science and technology.

Rajiv S. Mishra
University of North Texas
August 25, 2016

CHAPTER 1

Friction Stir Welding

Before the invention of friction stir welding, shortcomings of fusion welding of precipitation strengthened 2XXX aluminum alloys in creating structurally viable joints could not be eliminated. This led to mechanical fastener based joining as the key process for joining aluminum in aerospace applications.

Friction stir welding is a solid-state joining process that was invented by Thomas et al. [1] at The Welding Institute (TWI, UK) in 1991. FSW quickly gained research and industrial attention around the world and has been most widely studied on aluminum alloys during its first decade of evolution. The ability of FSW to join two separate parts without melting is the most significant feature. FSW has been most successful in aluminum-related industries as it can join any similar and dissimilar aluminum alloys with workpiece thickness ranging from 1-mm thin sheets to as thick as 75-mm thick blocks. More recently, other alloy systems, such as magnesium-, ferrous-, titanium-, copper-, nickel-based alloys, have also been subjected to FSW or its derivative processes [2–5]. Also, the ability of FSW to join dissimilar metals (alloys of different metals) in various configurations has been demonstrated successfully [6,7].

The basic principle of FSW is fairly simple. In this process, a non-consumable rotating tool, made of material stronger than workpiece, with a larger diameter shoulder and a pin, plunges into the workpiece to a preprogrammed depth. Plunging of rotating tool into the workpiece produces frictional heat due to the interaction of tool shoulder and workpiece. Another contribution to the heat input comes from adiabatic heat produced during plastic deformation of workpiece material around the rotating tool pin. Plastic deformation at high temperature leads to the softening of material around the pin. The softened material moves around with the rotation of tool pin which then traverses along a joint line and completes the weld. The larger diameter of tool shoulder helps in containing the hot material which can

Friction Stir Welding of 2XXX Aluminum Alloys Including Al–Li Alloys.
DOI: http://dx.doi.org/10.1016/B978-0-12-805368-3.00001-7

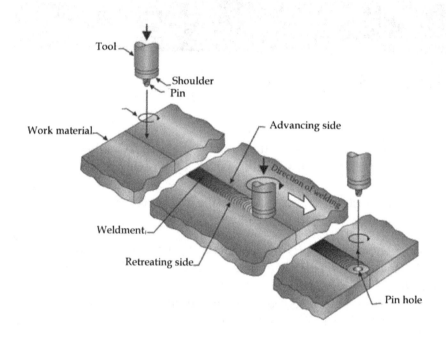

Figure 1.1 Schematic showing operational details of FSW process [10]. Source: (© 2012 Published in J.T. Khairuddin, I.P. Almanar, J. Abdullah, Z. Hussain, Principles and Thermo-Mechanical Model of Friction Stir Welding, INTECH Open Access Publisher, Croatia, 2012 under CC BY 3.0 license. Available from: http://dx.doi.org/10.5772/50156).

otherwise flow out easily to form flash and may lead to loss of material and defective weld. Schematic shown in Fig. 1.1 illustrates the process for welding of a workpiece in the butt configuration. During FSW, a solid-state joining process, melting does not take place. Thus, this process has inherent advantages and avoids the possibility of common defects like segregation, dendritic structure, and hot cracking, porosity formation associated with fusion-based welding techniques [3—5,8,9]. Dissimilar material can also be efficiently joined by using this process [3,5,6]. FSW has been used for many structural applications in aerospace and automobile industries [3]. Key benefits of FSW process over fusion welding techniques are listed in Table 1.1. FSW is also a green process as the energy consumption during this process is usually between 2% and 5% of energy consumption in arc-based welding [3,5]. Another benefit of this process is that the tool is nonconsumable and thus the requirement of filler material is completely avoided.

Although exit hole at the end of a conventionally made FSW is very commonly cited as a disadvantage of FSW, but in practice, there

Table 1.1 Key Benefits of FSW Process Over Conventional Fusion Welding Techniques [5]		
Advantages		Disadvantages
Metallurgical	Process Related	
No melting	No shielding gas	Huge process forces—special clamping required
No solidification cracking	No filler material	
No gas porosities	No harmful emissions	Exit hole at the end of the weld
No loss of alloying element	No work (arc) hazard	
Low distortion	Low workpiece cleaning	
Weld all aluminum alloys	No postweld milling required	
Excellent repeatability		
Nonconsumable tool		

Figure 1.2 Development of space shuttle external liquid hydrogen tank using friction stir welding, at the Marshall Space Flight Center (MSFC) NASA [4]. Source: Reproduced with permission of ASM INTERNATIONAL.

exist many possible solutions such as refilling probe [11] and other refill methods. Additionally, many design solutions are possible including a run-off tab so that can be later machined off.

As mentioned earlier, FSW has already been adopted and implemented in the industry. In case of joining aluminum, technology readiness level for FSW is high and it has been used by organizations like NASA, Boeing, Lockheed Martin, and Eclipse Aviation. NASA utilized FSW to build 2195 Al−Li alloy-based liquid hydrogen external tank of space shuttle at the Marshall Space Flight Center (MSFC) (Fig. 1.2) [4]. Eclipse Aviation implemented FSW in manufacturing of

Figure 1.3 Robotic friction stir welding (FSW) equipment used by Eclipse Aviation during manufacturing of the Eclipse 500 aircraft [4]. Source: Reproduced with permission of ASM INTERNATIONAL.

the Eclipse 500 business class jet (Fig. 1.3) [4]. Eclipse, with technical assistance from TWI, made extensive use friction stir welding and replaced approximately 7300 rivets per airframe [12].

1.1 FRICTION STIR WELDING OF ALUMINUM ALLOYS

Friction stir welding of aluminum alloys has been conducted for the past 20 years. A nonconsumable tool with a pin is made to traverse through the joint line of the material to be joined. Heat is produced due to friction between tool shoulder and workpiece which softens the material around the pin and produces a few microstructural changes in and around weld zone. Resultant microstructure of the weld cross-section, the size and distribution of the precipitates in the case of precipitation strengthened alloys, mechanical properties of the weldment, and other microstructural features are dependent on the welding parameters, tool dimensions, initial microstructure of workpiece material, and temperature distribution during welding [3,5,9].

Figure 1.4 Weld cross-section showing various metallurgical zones developed during FSW of an aluminum alloy [13].
Source: Reprinted with permission from Elsevier.

1.2 MICROSTRUCTURAL EVOLUTION IN FSW OF PRECIPITATION STRENGTHENED ALUMINUM ALLOYS

Typically, FSW of an aluminum alloy results in three different micro-structural zones as shown in Fig. 1.4: (1) weld nugget (WN), (2) thermo-mechnical affected zone (TMAZ), and (3) heat-affected zone (HAZ).

1.2.1 Weld Nugget

Plastic deformation and frictional heat generated during FSW results in transformation of parent material into equiaxed fine-grained recrystallized microstructure in the WN. Dislocation density in this zone in case of aluminum alloys is usually low [5]. Grain size in nugget usually ranges in 2–15 μm depending on FSW parameters, tool geometry, composition of material, and external cooling used [5]. In aluminum alloys, WN experiences a peak temperature in the range of 400–550°C during FSW [3,5]. In precipitation strengthened aluminum alloys, such high temperature can lead to coarsening, dissolution, and reprecipitation of precipitates depending on thermal stability of precipitate, alloy chemistry, external cooling, and peak temperature in WN [9].

1.2.2 Thermomechanical Affected Zone

Zone between parent material and nugget having experienced both temperature and plastic deformation during FSW is the TMAZ [4,5]. Partial recrystallization is generally observed in TMAZ [4,5]. Dislocation density is very high in TMAZ due to high plastic deformation which leads to formation of subgrain boundaries [5]. Dissolution of strengthening precipitates might occur in this zone during welding. Extent of dissolution varies depending on the thermal cycle experienced by TMAZ and the type of alloy.

1.2.3 Heat-Affected Zone

The zone separating TMAZ and base material is the HAZ. This zone does not undergo any plastic deformation and experiences thermal cycle. For a given welding parameter, formation of HAZ depends upon the alloy being welded. Generally, an area away from WN which experiences a thermal cycle in the range of 250—350°C during welding is observed as HAZ [4,5]. HAZ is characterized by the coarsening of strengthening precipitates and the precipitate-free zone formation due to high temperature exposure [5,9]. Use of external cooling medium can reduce the extent of formation of HAZ.

Usually FSW results in asymmetric weld cross-section. The side where tangential velocity of the tool is along the welding direction, is called the advancing side. The other side, where tangential velocity of the tool is opposite to the welding direction is called retreating side.

1.3 PROCESS VARIABLES

FSW is a complex process comprising material movement and plastic deformation of material at relatively high temperature. For a particular material, the resultant microstructure, mechanical, and corrosion properties of the joint is greatly influenced by many process-related external factors such as tool geometry, welding parameters, and joint configuration. A brief overview of these factors is presented in this section.

1.3.1 FSW Tools

In FSW process, unlike fusion welding techniques, external heat source is not required to generate heat during the process. FSW tool generates the entire heat and thus is the most influential factor in joining using FSW. As stated earlier, the FSW tool has two main components, shoulder and pin (also referred as probe), which are critically important in carrying out FSW. During the initial stage of plunging, the heat is generated due to friction between workpiece and pin. Once the shoulder touches the workpiece, it becomes the major contributor of heat generation due to the friction between larger surface area of shoulder and workpiece. The tool traverse begins at this stage and the tool pin drives the material movement around the pin to complete the joint. Tool pin also contributes to the heat content, generated due to the intense plastic deformation of the material.

Since FSW is a high temperature process involving high process forces acting on tool during welding, the tool material is an important consideration. Tool material used in manufacturing FSW tools is harder compared to workpiece material and must possess good high temperature properties. In FSW of aluminum, hardened tool steel (H13, D2) based tools are generally suitable for welding workpiece thickness less than 12 mm. For welding thicker sections and improvement in tool life, cobalt-based ultrahigh strength MP159 from the superalloy family is also utilized. Low cost of tool steel-based nonconsumable FSW tools adds another benefit to FSW of aluminum alloys.

Pin of FSW tool produces adiabatic heat as an outcome of plastic deformation of material around it. The pin used in the early days of FSW was essentially plain or threaded cylindrical with flat or round bottom (see Fig. 1.5). Threads on pin are designed to transport the material toward the bottom of the pin which ensures better material compaction and sound welds. Pin design progressively changed from cylindrical to tapered geometry (see Fig. 1.5). Tapered pins experience lower transverse forces due to lesser total surface area as compared to cylindrical pin. Consequently, the strongest part of the pin, the root of the tapered pin (the point where pin meets shoulder), takes the largest moment force. Tapered pins can be manufactured with either threaded

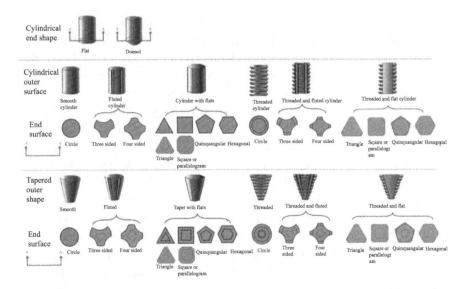

Figure 1.5 Schematic diagrams showing various features and shapes of pin of tools used in friction stir welding [14].
Source: Reprinted with permission from Taylor & Francis.

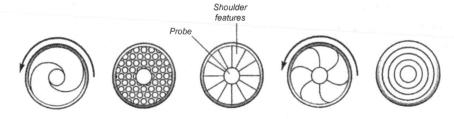

Figure 1.6 Various types of tool shoulder geometries [5]. Source: Reprinted with permission from Elsevier.

or step-spiral features. Step-spiral is easy to machine as compared to threads and considered better transporter of material as compared to the recesses of threads. Addition of other features such as flats were also experimented and found helpful in reducing process forces. In a comprehensive review about FSW tools, Zhang et al. [14] showed the various types of pin design of a FSW tool as shown in Fig. 1.5. TWI's efforts in tool development lead to a pin design known as Whorl pin, which reduces the volume of displaced material by 60% as compared to cylindrical pin of same diameter [4,14]. The important difference between a tapered pin with threads and the Whorl pin is the design of the helical ridges on the pin surface of the Whorl pin. TWI also produced the MX Triflute pin, which is a refinement of the Whorl pin containing three flutes machined into the helical ridges [4,5].

Tool shoulder of a FSW tool is designed to generate heat to soften the material and to produce enough downward forging force for material consolidation during welding. Concavity was the first design element to be utilized in FSW tool shoulder (see Fig. 1.6). The main aim was to contain the material inside the weld zone to produce defect-free weld. With further evolution of FSW, many design features were developed for tool shoulder. These features consist of scrolls, knurling, grooves, and concentric circles (see Fig. 1.6) machined with or without concavity. Another type of feature is convex shoulder with scrolls spiraling inwards with respect to intended rotation of the tool (see Fig. 1.7A). The benefit of this type of convex shoulder is that the entire shoulder of the tool need not be engaged. The shoulder can be only partially engaged with the workpiece. A sound weld is produced when any fraction of the scroll is engaged with the workpiece, since the scrolls are designed to provide a push to the moving material toward the center (weld zone). This design allows for a greater flexibility and can be used in joining two workpieces with different thickness. More

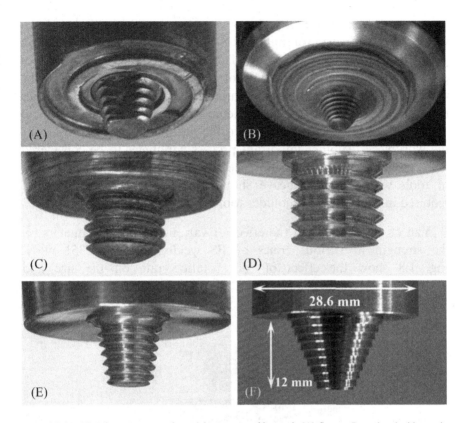

Figure 1.7 (A−F) A few most commonly used friction stir welding tools [4]. Source: Reproduced with permission of ASM INTERNATIONAL.

importantly, it can accommodate minor thickness variation in sheets/plates in production environment where lot-to-lot variation is expected. Fig. 1.7 shows some common tools used in FSW of aluminum alloys.

In FSW, much care is needed when through-thickness welds are required to be made. Although pin length is kept very close to the thickness of the workpiece, it is very difficult to create a perfect through-thickness joint. For such scenario, TWI developed a bobbin tool consisting of two shoulders, one on the top and other on the bottom surface of the workpiece, which are coupled by a pin fully engaged into the workpiece to produce through-thickness weld.

1.4 FSW PARAMETERS

In FSW, the two most important parameters are tool rotation rate and tool traverse speed along the line of joint. In FSW community, tool rotation rate is usually reported in revolutions per minute and tool traverse speed in inches (or millimeters) per minute. The tool rotation produces the stirring and mixing of material around the rotating tool pin and the traverse of tool transports the softened material from the front to the back of the pin and completes the welding process. Higher tool rotation rates generate higher temperature due to higher frictional heating. Tool traverse speeds govern the heating and cooling rates for the workpiece. Another important process parameter to be considered is the angle of the tool tilt with respect to the workpiece surface in case of tools with flat or concave shoulders. However, tool tilt is not required in case of scroll shoulder tools.

Yan et al. [15] studied the effect of various process parameters on the strength in various zones of FS welded AA2524-T351 alloy. Fig. 1.8 show the effect of tool rotation rate on the size and

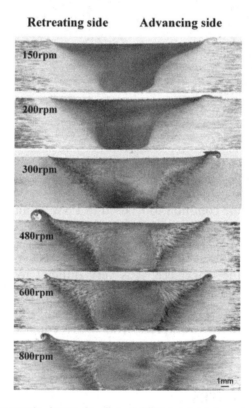

Figure 1.8 Etched macrographs showing the effect of different tool rotation rate on weld nugget [15].
Source: Reprinted with permission from Taylor & Francis.

macrostructure of WN of FS welded AA2524-T351 alloy. Yan et al. [15] showed that with higher tool rotation rate, the width of WN increases. This is due to the fact that the area of recrystallized region increases with increase in temperature at higher tool rotation rate.

1.5 FSW JOINT CONFIGURATION

During the initial days of FSW, majority of the joints made were in butt configuration. Though the most convenient joint configurations for FSW are butt and lap joints, other configurations such as T joints, double T joints, edge butt joints are also possible. A schematic of possible joint configuration is shown in Fig. 1.9. Recently, Palanivel et al. [16,17] successfully demonstrated the feasibility of FSW to additively build the structure of aluminum and magnesium alloys using multiple lap configuration.

Majority of the commercial FSW applications involve simple butt joint configurations. More complex designs like sections containing T joints and corner welds are rarely considered. Baumann [18] (from The Boeing Company), in collaboration with TWI, developed corner angle welding for vertical stiffener joining using FSW. They used a squared stationary shoulder to consolidate the weld. A conceptual schematic of the process, tool pin, and tool shoulder is shown in Fig. 1.10.

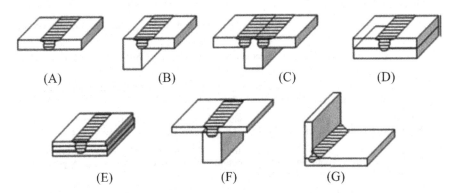

Figure 1.9 Joint configurations for friction stir welding: (A) square butt, (B) edge butt, (C) T butt joint, (D) lap joint, (E) multiple lap joint, (F) T lap joint, and (G) fillet joint [5]. Source: Reprinted with permission from Elsevier.

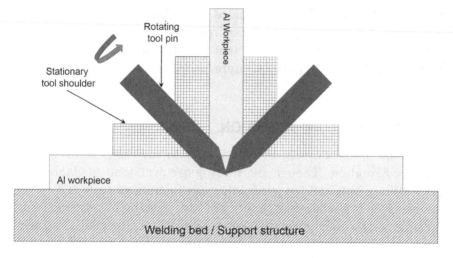

Figure 1.10 Schematic showing a setup to make corner welds using FSW.

REFERENCES

[1] W. Thomas, E. Nicholas, J. Needham, M. Murch, P. Templesmith, C. Dawes. International patent application no. (1991).

[2] R. Nandan, T. DebRoy, H. Bhadeshia, Recent advances in friction-stir welding–process, weldment structure and properties, Prog. Mater. Sci. 53 (2008) 980–1023.

[3] R.S. Mishra, P.S. De, N. Kumar, Friction Stir Welding and Processing: Science and Engineering, Springer, USA, 2014.

[4] R.S. Mishra, M.W. Mahoney, Friction Stir Welding and Processing, ASM International, Ohio, 2007.

[5] R.S. Mishra, Z. Ma, Friction stir welding and processing, Mater. Sci. Eng. 50 (2005) 1–78.

[6] N. Kumar, R.S. Mishra, W. Yuan, Friction stir welding of dissimilar alloys and materials, Butterworth-Heinemann, UK, 2015.

[7] Y. Hovanski, G.J. Grant, S. Jana, K.F. Mattlin, Friction stir welding tool and process for welding dissimilar materials. (2013).

[8] R.S. Mishra, P.S. De, N. Kumar, Fundamentals of the Friction Stir Process, Springer, USA, 2014.

[9] P.S. De, R.S. Mishra, Friction stir welding of precipitation strengthened aluminium alloys: scope and challenges, Sci. Technol. Weld. Joining 16 (2011) 343–347.

[10] J.T. Khairuddin, I.P. Almanar, J. Abdullah, Z. Hussain, Principles and Thermo-Mechanical Model of Friction Stir Welding, INTECH Open Access Publisher, Croatia, 2012.

[11] Y. Uematsu, K. Tokaji, Y. Tozaki, T. Kurita, S. Murata, Effect of re-filling probe hole on tensile failure and fatigue behaviour of friction stir spot welded joints in Al–Mg–Si alloy, Int. J. Fatigue. 30 (2008) 1956–1966.

[12] P. Threadgill, A. Leonard, H. Shercliff, P. Withers, Friction stir welding of aluminium alloys, Int. Mater. Rev. 54 (2009) 49–93.

[13] H. Sidhar, N.Y. Martinez, R.S. Mishra, J. Silvanus, Friction stir welding of Al–Mg–Li 1424 alloy, Mater. Des 106 (2016) 146–152.

[14] Y. Zhang, X. Cao, S. Larose, P. Wanjara, Review of tools for friction stir welding and processing, Can. Metall. Q 51 (2012) 250–261.

[15] J Yan, MA Sutton, AP Reynolds. Process–structure–property relationships for nugget and heat affected zone regions of AA2524–T351 friction stir welds, Sci. Technol. Weld. Joining. (2013).

[16] S. Palanivel, H. Sidhar, R. Mishra, Friction stir additive manufacturing: route to high structural performance, JOM 67 (2015) 616–621.

[17] S. Palanivel, P. Nelaturu, B. Glass, R. Mishra, Friction stir additive manufacturing for high structural performance through microstructural control in an Mg based WE43 alloy, Mater. Des 65 (2015) 934–952.

[18] J.A. Baumann, Production of Energy Efficient Preform Structures (PEEPS). (2012).

CHAPTER 2

Physical Metallurgy of 2XXX Aluminum Alloys

2.1 INTRODUCTION

Rising demand for fuel economy, reduction in carbon footprint, and improved structural efficiency are pushing factors for light weighting in automotive and aerospace industries. Aluminum is the most abundant metal on earth [1] and is the second lightest metal after magnesium, commercially used in alloy form. Combination of various mechanical, physical, thermal, corrosion properties, and economics makes aluminum alloys a key candidate in aerospace applications [2]. Since the early days of commercial aircraft manufacturing, aluminum has been the main material for primary aircraft structures [3]. Although, recently, polymer matrix composite materials are showing promising results and are being used significantly in some cases (the Boeing 787 Dreamliner) [4,5], aluminum alloys are still considered as primary material for airframe construction and contribute 60% of structural weight in case of Airbus A380 passenger aircraft [4].

Aluminum alloys are produced in both cast and wrought forms. Wrought aluminum alloys are classified based on the main alloying element. Table 2.1 shows the standard four-digit designation developed by the Aluminum Association for wrought aluminum alloys [6]. This designation is followed by majority of the countries around the world and is referred as International Alloy Designation System (IADS). Wrought aluminum alloys can be further classified into heat treatable and nonheat treatable alloys. The 1XXX, 3XXX, and 5XXX alloys are nonheat treatable and strengthened by work hardening [2]. The 2XXX, 6XXX, 7XXX, and some of 8XXX belong to the heat treatable (age hardenable or precipitation-strengthened) aluminum alloys category [2]. These alloys possess highest specific strength in aluminum alloys. Most of the heat treatable alloys are highly alloyed with other elements to achieve high strength. 9XXX alloy series are still unused and kept for new type of aluminum alloys in future.

Friction Stir Welding of 2XXX Aluminum Alloys Including Al–Li Alloys.
DOI: http://dx.doi.org/10.1016/B978-0-12-805368-3.00002-9

Table 2.1 Standard Designation for Wrought Aluminum Alloys	
Standard Designation	**Main Alloying Elements**
1XXX	None (>99% Aluminum)
2XXX	Copper
3XXX	Manganese
4XXX	Silicon
5XXX	Magnesium
6XXX	Magnesium and silicon
7XXX	Zinc
8XXX	Others
9XXX	Unused

The majority of wrought aluminum alloys gain strength from precipitation hardening. Al 2XXX, 6XXX, and 7XXX series alloys belong to this category. The strength achieved from precipitation process is highly dependent on the type, quantity, and mutual interaction of alloying elements. Variation in alloy chemistry leads to a wide spectrum of tensile strength in aluminum alloys—as high as 300 MPa in peak aged 6XXX alloys to 700 MPa in 7XXX alloys. 2XXX and 7XXX series alloys make the strongest aluminum alloys produced. Due to excellent combination of low density and high strength, 2XXX and 7XXX alloys are most widely used metallic systems in the aerospace industry where specific strength is an important factor.

Aluminum 2XXX alloys have been associated with aircraft industry from the inception of commercial aircraft manufacturing. The first alloy which was considered in aircraft design was Al—Cu—Mg alloy popularly known as Duralumin [7]. Damage tolerant alloy 2024 was incorporated in aircraft design in the 1950s. Since then, 2XXX alloys have grown as one of the two major alloy systems used in the aerospace industry [3]. Similarly, other Al—Cu-based alloys were developed in the last century. Lithium bearing Al—Cu alloys have been developed due to excellent combination of low density and high strength offered by these alloys.

The first commercial Al—Li alloy, "Scleron," was produced in Germany in 1924 with a nominal composition of Al—12Zn—3Cu—0.6Mn—0.1Li [8]. Al—Li alloys have gained popularity from the fact that addition of each weight percentage of lithium, the lightest metallic element, enhances the elastic modulus by 6% and reduces the density of resultant binary aluminum alloy by 3% [9,10]. Except beryllium, which is toxic, lithium is the only metal that decreases the density and improves the elastic modulus of aluminum [10] (Fig. 2.1). Lower density and higher modulus provide high stiffness (elastic modulus/density) and high specific strength (strength/density) which enable large weight savings in airplane structures and improve the fuel efficiency. Also,

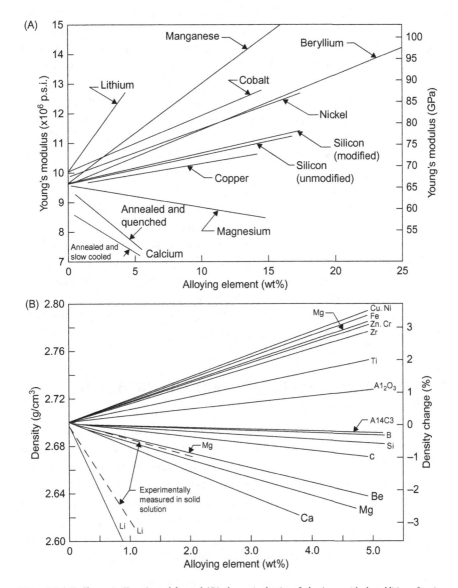

Figure 2.1 (A) Change in Young's modulus and (B) change in density of aluminum with the addition of various elements and compounds [10]. Source: Reprinted with permission from Springer.

modern (third-generation) Al–Li alloys possess excellent mechanical properties (high tensile strength, improved high cycle fatigue, and fatigue crack growth resistance) comparable to incumbent aerospace 7XXX and 2XXX alloys [8]. After short stint and withdrawal of Scleron, it took nearly four decades when Alcoa developed Al–Li alloy 2020 with nominal composition of Al–4.5Cu–1.1Li–0.5Mn–0.2Cd [8]. Alloy 2020 was particularly impressive due to high strength at high temperatures (150–200°C). Alloy 2020 was used in wing and tail structure of RA 5C Vigilante aircraft for two decades before it was discontinued due to ductility and production problems [8]. However, no failures were reported during its use for a period of 20 years [8]. It began the era of development of Al–Li alloys around the globe. It was followed by the development and use of much lighter and moderate strength 1420 Al–Mg–Li based alloys in USSR in 1960s [5]. Fridlyander pioneered the development of Al–Li alloys in Soviet Union and used in Soviet aircraft structures [5,8,11–13]. In recent times, Al–Li alloys have regained the interest and are being developed rapidly [4,8]. Replacement of 2219 (Al–Cu alloy) with 2195 (Al–Cu–Li alloy), which is 30% stronger and 5% less dense than 2219, in manufacturing of external fuel tank of space shuttle by NASA [14,15] shows the potential of the Al–Cu–Li–X family of alloys in aerospace and aviation sector. There are numerous other industrial applications of Al–Cu–Li–X alloys which includes Airbus A380 (2196 alloy), F16 aircraft (2297), Boeing 787 dreamliner (2099 and 2199), and Agusta Westland helicopter [4,8].

2.2 WELDABILITY OF PRECIPITATION-STRENGTHENED ALUMINUM ALLOYS

Generally, 2XXX aluminum alloys are considered unweldable using conventional fusion-based welding techniques [3,4,16–20]. Volatile nature of Mg, Zn, and Li makes joining of these alloys difficult using laser-based welding technology [16,20]. Hydrogen porosity, hot cracking, stress corrosion cracking are the main issues associated with fusion welding of precipitation-strengthened aluminum alloys [16,17,19,20]. Fig. 2.2 shows two examples of hydrogen porosity and keyhole porosity formed in laser beam-welded Al–Li alloy.

The main cause of porosity formation during fusion welding of Al–Li and 2XXX alloys is hydrogen contamination. Solubility of

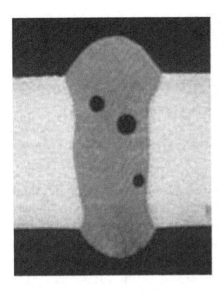 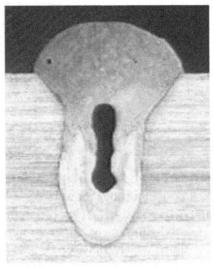

Figure 2.2 Examples of porosity formation in laser beam welding of Al–Li alloys [16]. Source: Reprinted with permission from Elsevier.

hydrogen in molten aluminum is considerably high and reduces by a great extent below solidus temperature. Therefore, when weld metal solidifies, hydrogen gets trapped into weld pool and forms hydrogen bubbles. Bubbles end up forming porosity if they cannot escape the molten weld pool before solidification leading to the formation of porosities as shown in Fig. 2.2. Although shielding gases are used in most of the fusion welding techniques, it is extremely difficult to completely eliminate hydrogen from moisture around the weld pool due to its high diffusivity.

Hot cracking is another persistent issue in fusion welding of Al–Li and 2XXX alloys [16,20]. It usually occurs in weld nugget during solidification. It occurs due to the chemical segregation of low melting eutectic along the grain boundaries during final stages of solidification [16,17,20,21]. Cracking occurs when the stresses developed due to thermal shrinkage across the neighboring grains exceed the strength of solidifying metal. Depending upon thermal cycle and alloy chemistry, sometimes solidification cracking is also observed in parent material (HAZ) away from weld pool as shown in Fig. 2.3. In 2XXX alloys, high alloying content severely increases the susceptibility to hot cracking, mainly due to Al–Cu–Mg eutectic.

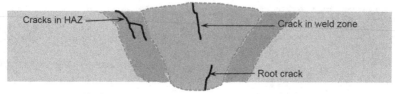

Weld cross section of fusion weld of 2XXX alloy

Figure 2.3 Schematic showing examples of hot cracking in weld zone and heat-affected zone during fusion welding of aluminum alloys.

Apart from the above-mentioned issues, fusion-based welding techniques consume huge power, generate toxic gases, and require highest safety standards [17,19]. Strength and fatigue life of fusion-welded structural component can be compromised due to the presence of microporosities which may go undetected during nondestructive testing. Solid-state joining techniques may solve most of the issues associated with fusion welding of precipitation-strengthened aluminum alloys.

2.3 CLASSIFICATION OF 2XXX ALUMINUM ALLOYS

2XXX aluminum alloys can be considered as one of the most complex precipitation-strengthened systems. The intriguing complexity of 2XXX alloys can be understood from the fact that it took nearly a century to ascertain the origin of initial aging response in Al–Cu–Mg-based alloys [22,23], the same alloy system which led to the discovery of precipitation hardening in 1906 [24,25]. Similarly, several structures for various precipitates have been proposed. Wang and Starink [24] provided a comprehensive review of precipitates and other second phase particles in 2XXX alloys. 2XXX alloys are precipitation-strengthened alloys and hence a detailed insight into the precipitates involved in this system is of utmost importance. In an alloy, precipitation sequence and the nature of the precipitates is governed by the alloying elements present, their absolute, and relative (ratios) concentrations. Other than Cu, Mg and Li are most common alloying elements in 2XXX alloys. Minor alloying elements, such as Zr, Ag, Mn, and Sc, are also added to enhance certain targeted mechanical properties. Therefore, it is important to further identify and discuss various alloy subcategories under 2XXX family. The effect of various alloying elements and their relative (ratio) concentrations on the precipitation will also be discussed later.

Addition of Cu as main alloying element defines the basis of 2XXX series of aluminum alloys. Further addition of other alloying elements, such as Mg and Li, and addition of Ag, Zn, Sc, Cr, Zr, Mn, Ti in smaller quantities are utilized throughout the 2XXX alloys. Based on the alloy chemistry, 2XXX series of aluminum alloys can be classified into three broad categories: Al–Cu alloys, Al–Cu–Mg alloys, and Al–Cu–Li alloys. Alloy composition of a few commercially used 2XXX alloys is listed in Table 2.2.

2.4 PHYSICAL METALLURGY OF 2XXX ALUMINUM ALLOYS

Precipitation hardening is the main strengthening phenomenon in 2XXX aluminum alloys. Precipitates form during aging at ambient or elevated temperature. The basic criterion for an alloy to be categorized as precipitation-strengthened alloy is a decrease in solid solubility of alloying elements with decrease in temperature. It can be explained with the classic example of precipitation in Al–Cu system as shown in Fig. 2.4. The thermal treatment required in this process starts with solution heat treatment by soaking the alloy at sufficiently high temperature for long enough time to form solid solution (step 1 in Fig. 2.4). It is followed by rapid quenching to room temperature to form supersaturated solid solution (step 2 in Fig. 2.4). Precipitation of secondary phases occurs at room temperature (natural aging) or elevated temperature (artificial aging, step 3 in Fig. 2.4) due to the decrease in solid solubility of alloying elements at lower temperatures. Artificial aging can be a single- or multistep thermal treatment depending upon alloy chemistry and targeted property.

2.4.1 Physical Metallurgy of Al–Cu Alloys

The phenomenon of precipitation hardening was first discovered in an Al–4Cu–0.6Mg (wt%) alloy [24]. Al–Cu system is the most studied alloy system to understand age hardening. The precipitation in Al–Cu alloys during aging is highly dependent on the aging temperature and level of supersaturation. The precipitation sequence in Al–Cu alloys system has been widely studied and can be described by the following sequence:

$$\text{Supersaturated Solid Solution(SSSS)} \rightarrow \text{GP zones} \rightarrow \theta'' \rightarrow \theta' \rightarrow \theta \quad (2.1)$$

GP zones are clusters of Cu atoms formed during initial stages of artificial aging or prolonged natural aging on $\{1\,0\,0\}$ matrix planes.

Table 2.2 Alloy Composition Limits of a Few Commercially Used 2XXX Aluminum Alloys [6]

Alloys	Cu	Mg	Li	Ag	Mn	Zr	Zn	Cr	Si	Fe
2014	3.9–5.0	0.2–0.8	–	–	0.4–1.2	–	0.25	0.10	0.5–1.2	0.7
2017	3.5–4.5	0.4–0.8	–	–	0.4–1.0	–	0.25	0.10	0.2–0.8	0.7
2219	5.8–6.8	0.02	–	–	0.2–0.4	0.10–0.25	0.10	–	0.2	0.3
2024	3.8–4.9	1.2–1.8	–	–	0.3–0.9	–	0.25	0.10	0.5	0.5
2224	3.8–4.4	1.2–1.8	–	–	0.3–0.9	–	0.25	0.10	0.12	0.15
2524	4.0–4.5	1.2–1.6	–	–	0.45–0.7	–	0.15	0.05	0.06	0.12
2139	4.5–5.5	0.2–0.8	–	0.15–0.6	0.2–0.6	–	0.25	0.05	0.10	0.15
2040	4.8–5.4	0.7–1.1	–	0.4–0.7	0.45–0.8	0.08–0.15	0.25	–	0.08	0.10
2090	2.4–3.0	0.25	1.9–2.6	–	0.05	0.08–0.15	0.1	0.05	0.1	0.12
2195	3.7–4.3	0.25–0.8	0.8–1.2	0.25–0.6	0.25	0.08–0.16	0.25	–	0.12	0.15
2199	2.3–2.9	0.05–0.4	1.4–1.8	–	0.1–0.5	0.05–0.12	0.2–0.9	–	0.05	0.07
2050	3.2–3.9	0.2–0.6	0.7–1.3	0.2–0.7	0.2–0.5	0.06–0.14	0.25	0.05	0.08	0.10
2060	3.4–4.5	0.6–1.1	0.6–0.9	0.05–0.5	0.1–0.5	0.05–0.15	0.3–0.5	–	0.07	0.07

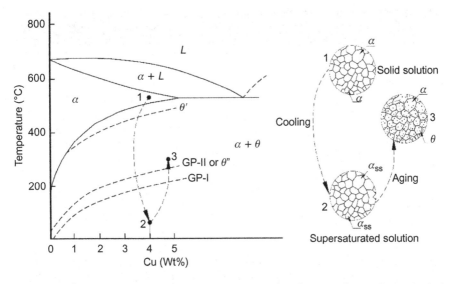

Figure 2.4 Aluminum-rich binary phase diagram of Al–Cu system with a schematic to illustrate the steps involved in aging treatment of precipitation-strengthened aluminum alloys [26]. Source: Reprinted with permission from Taylor & Francis.

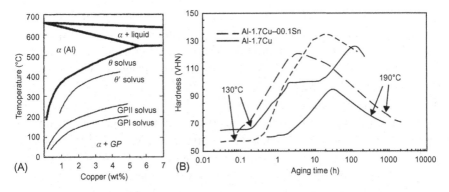

Figure 2.5 (A) Al–Cu binary phase diagram showing solvus boundaries of various phases in Al–Cu alloy system [28] and (B) hardness time plot for Al–1.7Cu alloy with and without the addition of Sn [25]. Source: Reprinted with permission from Elsevier.

GP zones evolve from two-dimensional disc shapes to three-dimensional zones with aging time. GP zones in Al–Cu alloy system are stable at relatively low temperatures and tend to redissolve into aluminum matrix when subjected to elevated temperature aging. GP zones, θ'' and θ', are metastable phases, whereas θ is the equilibrium phase of Al–Cu alloy system. GP zones and θ'' have similar composition of 25–45% Cu. The solvus of metastable and equilibrium phases of Al–Cu system is shown in Fig. 2.5A [24]. Body-centered tetragonal

Table 2.3 Structure, Space Group, Lattice Parameters, and Compositional Details of Key Phases Observed in 2XXX Aluminum Alloys

Phases	Structures	Space Groups	Lattice Parameters (nm)			Composition	References
			a	b	c		
θ'	Tetragonal	I4m2	0.404	–	0.58	Al_2Cu	[8,25]
θ	Tetragonal	I4/mcm	0.607	–	0.487	Al_2Cu	[28]
S	Orthorhombic	Cmcm	0.4	0.923	0.714	Al_2CuMg	[8,25]
Ω	Orthorhombic	Fmmm	0.496	0.859	0.848	Al_2Cu	[33]
T	Cubic	T^h_5	1.425	–	–	Al_6CuMg_4	[28]
T_1	Hexagonal	P6/mmm	0.497	–	0.935	Al_2CuLi	[8]
δ	Cubic	B32	0.638	–	–	$AlLi$	[8]
δ'	Cubic	$Pm\overline{3}m$	0.405	–	–	Al_3Li	[8]

structure and space group I4/mcm is the widely accepted structure of θ which was first proposed by Silcock et al. [24,27]. Ringer and Hono [28] proposed similar structure for equilibrium phase θ with different lattice parameters as presented in Table 2.3.

Fig. 2.5B shows hardness evolution with time in an Al–1.7Cu (wt%) alloy aged at 130°C and 190°C. Aging at lower temperature (e.g., 130°C) results in two-stage hardening behavior, whereas aging at higher temperatures (190°C) shows one-stage hardening with continuous increase in hardness as shown in Fig. 2.5B [25]. This behavior is mainly due to the stability of GP zones during low temperature aging which then transforms into θ'' during second stage of hardening response. GP zones in Al–Cu alloys are considered stable at 130°C and begin to dissolve at temperature around 150°C. Aging at higher temperature bypasses the formation of GP zones and solid solution directly transforms into θ'' which can further transform into θ'. θ' precipitates act as the main strengthening phase in binary Al–Cu system. Aging for extended time period at high temperatures (overaging) results in the formation of coarse θ equilibrium precipitate at the expense of θ' precipitate. While θ' and θ both share the same composition of Al_2Cu and tetragonal structure, θ' is considered semicoherent with the aluminum matrix. The crystal structures of all the phases of Al–Cu alloy system are given in Table 2.3.

2.4.2 Physical Metallurgy of Al–Cu–Mg Alloys

The addition of Mg to Al–Cu system further enhances the strength of the alloy. Mg addition decreases the density of the alloy and increases

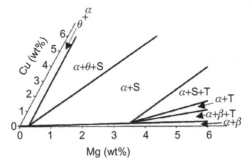

Figure 2.6 Isothermal section of ternary Al−Cu−Mg phase diagram at 200°C [24]. Source: Reprinted with permission from Taylor & Francis.

the strength via solid solution hardening and precipitation hardening. Compositional variation in the Al−Cu−Mg system results in different equilibrium phases. Most of the industrially used alloys of 2XXX series belong to this category. Al−Cu−Mg alloys can be subcategorized into four phase fields, i.e., $\alpha + \theta$, $\alpha + \theta + S$, $\alpha + S$, and $\alpha + S + T$, as shown in Fig. 2.6. Theoretically, a few more minor phase fields (Fig. 2.6) can also form but are not industrially utilized. A number of precipitates are possible in the Al−Cu−Mg alloy system; θ'', θ', and θ from the binary Al−Cu system, and Ω, S, T, and β. S'' and S' have been suggested to be the precursors of the S phase. The precipitation sequence for this alloy system is as follows:

$$\begin{aligned} SSSS \rightarrow Clusters \quad &\rightarrow \quad GP \text{ zones} \rightarrow \theta'' \rightarrow \theta' \rightarrow \theta \\ &\rightarrow \quad GPB \text{ zones} + S' \rightarrow S \quad\quad\quad (2.2) \\ &\rightarrow \quad \Omega \rightarrow \theta \end{aligned}$$

As evident from Fig. 2.6, Al−Cu−Mg alloys with medium to low Cu:Mg ratio ($\sim 0.5-2$), which lie in $\alpha + S$ phase field, demonstrate a two-stage hardening response. This behavior is usually observed during a typical T6 heat treatment between 110°C and 240°C [25,30]. During aging, a prolonged hardness plateau is observed between first-stage and second-stage hardening as shown in Fig. 2.7. The first stage of hardening is usually very quick and is over in less than 2 minutes of artificial aging. The extent of hardness increase in first-stage hardening depends on alloy chemistry and increases with increase in Mg content (Fig. 2.7) [30,31]. The second stage of hardening after plateau takes much longer time (from hours to days) and depends on the aging temperature [30].

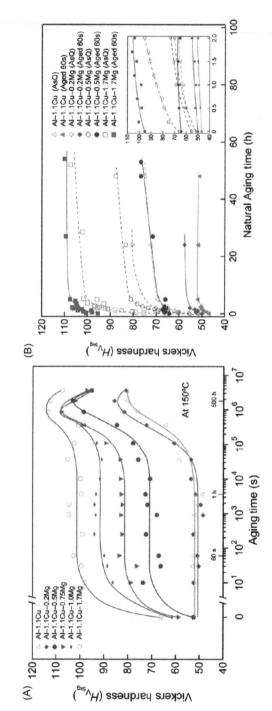

Figure 2.7 (A) Evolution of hardness in Al−1.1Cu−xMg (where x = 0, 0.2, 0.5, 0.75, 1.0, 1.7 at%) alloys aged at 150°C and (B) natural aging response of Al−1.1Cu−xMg (where x = 0, 0.2, 0.5, and 1.7 at%) alloys after solution treatment for 1 hour at 525°C and quenching, and also after artificial aging for 60 seconds at 150°C and further quenching [30].

Source: Reprinted with permission from Elsevier.

In Al−Cu−Mg alloys, Gunier−Preston−Bagaryatsky (GPB) zones were universally accepted as the reason for rapid hardening during first stage of hardening response, until in 1997, when Ringer et al. [22] using atom probe technique discovered that atomic level co-clusters of Cu and Mg were responsible for rapid hardening during the first stage. Cu−Mg co-cluster was found to be fully coherent with aluminum matrix and extremely small in size (<20 atoms per cluster). These clusters form on the quenched-in vacancies and dislocations, increase the strength by locking the dislocations. Size of Cu−Mg clusters grows during the hardness plateau [30]. A recent study by Sha et al. [32] on commercial 2024 alloy shows that the large Cu−Mg clusters with higher Mg concentration continuously transforms into GPB zones during aging. Sha et al. [32] also found that in peak aged condition, Cu−Mg clusters, GPB zones, and S (or S′) phases coexist. Wang and Starink [24] concluded in their review assessment that the second stage of hardening is mainly due to formation of S phase. The width of plateau in hardness evolution is observed to lower with increase in aging temperature for Al−Cu−Mg alloys in α + S phase field [24].

The alloys based on α + S + T phase field have rarely found industrial applications as these do not offer any advantage as compared to alloys in α + S phase field and offer similar strength levels [24,25].

Addition of Ag in trace quantities enhances the precipitation hardening in all Al−Cu−Mg alloys with high Cu:Mg ratio [24,25,28]. Ag addition has no effect on precipitation sequence or behavior of θ' precipitate. Aging of Al−Cu−Mg−Ag alloys of α + θ and α + θ + S phase fields results in the formation of Ω precipitate. Ω precipitates on {1 1 1} planes of aluminum matrix [24,28]. Composition of this phase is similar to θ phase (Al$_2$Cu) of the binary Al−Cu alloy system. Although many structures have been proposed for this phase, orthorhombic structure with Fmmm space group proposed by Knowles and Stobbs [33] is widely accepted [28]. Alloys containing Ω phase show improved high temperature strength and hence have attracted a significant research interest [28,34]. Overaging of these alloys result in the replacement of Ω precipitates with the equilibrium θ precipitate [34].

2.4.3 Physical Metallurgy of Al−Cu−Li Alloys

Al−Li alloys are another subcategory of 2XXX alloys. Al−Li alloys are also precipitation-strengthened alloys, i.e., strength can be

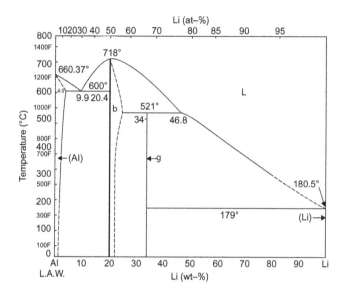

Figure 2.8 Binary phase diagram of Al–Li system [20]. Source: Reprinted with permission from Taylor & Francis.

increased by solution heat treatment and then aging. Strength of these alloys is dependent on the type, size, volume fraction, and distribution of precipitation in the matrix and at grain boundaries.

Lithium has high solubility in aluminum with a maximum of 16 at% (4 wt%) at 600°C as shown in Fig. 2.8. Binary Al–Li alloys are strengthened by precipitation of metastable, coherent, and spheroidal Al_3Li (δ') precipitate [35–37]. δ' is a $L1_2$ ordered phase and has very small misfit strain and interfacial energy of ≈ 10 mJ/m^2 [35,38]. At equilibrium, δ' transforms into stable δ (AlLi) and solid solution matrix of aluminum [10,35]. Hardening phase δ' may also be sheared by a moving dislocation which gives rise to slip planarity leading to poor ductility and hence considered as detrimental in Al–Li–X alloys [8]. Fig. 2.9 summarizes various strengthening precipitates possible in different Al–Li–X alloys.

On aging (below the solvus of δ') after a solution treatment, δ' precipitates in spheroidal form and the sequence of precipitation for a binary Al–Li can be described as [35]

$$SSSS \rightarrow \alpha + \delta' \rightarrow \alpha + \delta(\text{equilibrium})$$

δ is the equilibrium phase (AlLi) which precipitates out after dissolution of metastable δ' on overaging. Binary Al–Li-based alloys have no practical application.

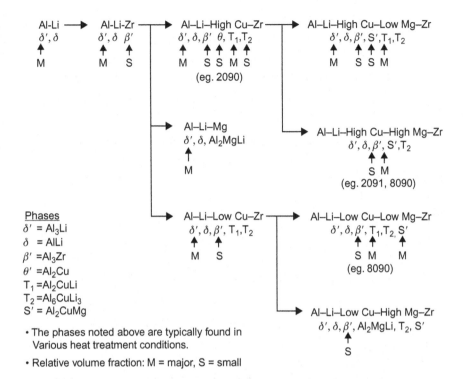

Figure 2.9 Flowchart showing possible phase in various binary, ternary, and quaternary Al–Li alloys [8].
Source: Reprinted with permission from Elsevier.

2.4.3.1 Al–Mg–Li System

A number of Al–Mg–Li alloys were developed and studied in great detail by Russian researchers at the Institute of Aviation Materials (VIAM), Moscow [5,11–13]. Variants of high Mg and Li content were developed such as 1420, 1421, and 1423. δ' is the main strengthening precipitate and T phase, also referred to as S_1 (cubic, Al_2LiMg), forms on overaging (stabilized) [5,8,12].

Noble and Thompson [35] suggested the precipitation sequence in the Al–Mg–Li system and is as follows:

$$SSSS \rightarrow \delta'(Al_3Li) \rightarrow \delta(AlLi) \quad \text{[for high Li/Mg ratio]}$$

or

$$SSSS \rightarrow \delta'(Al_3Li) \rightarrow T(Al_2MgLi) \quad \text{[for low Li/Mg ratio]}$$

Mg has shown to reduce the solubility of Li in aluminum and therefore increases the Al_3Li precipitation [10].

2.4.3.2 Al—Cu—Li System

Al—Cu—Li alloys contains all the phases present in Al—Cu and Al—Li systems, i.e., θ', θ (Al_2Cu), and δ' (Al_3Li). T_1 (hexagonal, Al_2CuLi) is also observed and is the main strengthening precipitate along with θ and δ' in the Al—Cu—Li system. T_1 forms as thin and long plate-type structures on the {1 1 1} of matrix planes [39—42].

Cu:Li ratio has a huge impact on the type and volume fraction of precipitates in Al—Cu—Li alloys. Jo and Hirano [43] studied the effect of Cu:Li ratio on the precipitation in Al—Cu—Li alloy and their findings are as follows:

- For high Cu:Li (>4), SSSS$\rightarrow$GP zones$\rightarrow\theta''\rightarrow\theta'$
- For Cu:Li $= 2.5-4.0$, SSSS$\rightarrow$GP zones$\rightarrow$GP zones $+\,\delta'\rightarrow\delta' + \theta'' + \theta'\rightarrow\delta' + T_1$
- For Cu:Li $= 1-2.5$, SSSS$\rightarrow$GP zones $+\,\delta'\rightarrow\delta' + \theta'\rightarrow\delta' + T_1\rightarrow T_1$
- For Cu:Li <1, SSSS$\rightarrow\delta' + T_1\rightarrow T_1$.

Decreus et al. [44] also carried out similar study on 2198 (Cu:Li $= 3.36$) and 2196 (Cu:Li $= 1.65$), though both the alloys are modern Al—Li alloys, and found results in accordance with Jo and Hirano [43]. 2198 was observed to show Cu clusters during natural aging whereas 2196 shows δ' precipitation. Cu also lowers the solubility of Li in aluminum and hence promotes greater precipitation of Li containing precipitates.

2.4.3.3 Al—Cu—Mg—Li—X System

Most of the industrially utilized and latest third-generation Al—Li alloys around the globe belong to this category. In many of these alloys, small amount of Ag (usually 0.25 wt%) is added to improve strength [45]. Alloy composition of some of the third-generation Al—Li alloys is listed in Table 2.2. Addition of minor alloying elements along with Li and Mg introduces improved strengthening in these alloys as compared to Al—Cu—Li alloys. Ag is known to stimulate the precipitation of T_1 phase [46]. Ag accommodates the misfit energy between precipitate and aluminum matrix by segregating on matrix—precipitate interface [46]. Mg added alone to the ternary Al—Cu—Li alloy promotes GP zone formation and precipitation of early θ' and T_1 in later stages of aging treatment [46].

2.4.4 Constituent Phases and Dispersoids

Size, distribution, morphology, and type of precipitates in 2XXX alloys control a major fraction of strength of the alloy. Other second phase particles also exist which can affect mechanical properties in many ways. Such phases can be categorized into constituent phases and dispersoids. Constituent phases, also sometimes referred as coarse intermetallic phases, form due to eutectic reaction during solidification [24]. Size of these particles generally ranges from several tens of microns to hundreds of microns. In structural applications, these phases are highly undesirable as they are detrimental to fatigue resistance and fracture toughness of the alloy. Size and distribution of constituent phases are often controlled via deformation during thermo-mechanical processing. Constituent phase can be further divided into soluble and insoluble phases. Mainly, constituent phases are particles containing impurity elements such Si and Fe which have low solid solubility in Al (Al−Mg) alloys. Fe and Si impurity content is not eliminated completely as it increases the production cost significantly and is only controlled below a certain level to achieve desired properties. Main alloying elements, such as Mg, Cu, and Li, can also result in the formation of Al_2Cu, Al_2CuMg, and Al_2CuLi type constituent phases. These phases are mostly soluble and thus are dissolved during homogenization process provided the solubility of the element in aluminum is high enough. Constituent phases $Al_{12}(Fe,Mn)_3Si$, $Al_6(Fe, Cu)$, Al_7Cu_2Fe, Mg_2Si, Al_2Cu, Al_2CuMg, Al_3Fe, Al_2CuLi, and Al_6CuLi_3 are generally observed in 2XXX alloys [2,8,24].

Size of constituent particles is highly dependent on solidification process and generally decreases with increase in solidification rate. Some constituent phases can also result in incipient melting which can severely affect the high temperature thermomechanical processing. The insoluble phases consisting Fe and Si are particularly detrimental for structural components where fatigue and fracture toughness are key design elements. Fig. 2.10 shows the effect of Si and Fe content on fracture toughness of 2×24 alloys tested in peak aged condition. These phases are highly incoherent in nature and do not provide any strength increment as evident from Fig. 2.10.

Dispersoids are formed during solidification or homogenization treatment of alloy ingot. Mn, Zr, and Cr are the key elements added

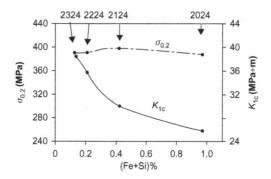

Figure 2.10 Effect of Fe and Si impurity contents on strength and fracture toughness of 2X24 series alloys aged at 190°C for 12 hours [24]. Source: Reprinted with permission from Taylor & Francis.

Table 2.4 A List of Constituent Phases Commonly Found in Various 2XXX Alloys [2,24]	
Alloy Types	**Constituent Phases**
2 × 24	Al_7Cu_2Fe, $Al_{12}(Fe,Mn)_3Si$, Al_2CuMg, Al_2Cu, $Al_6(Cu,Fe)$
2 × 19	Al_7Cu_2Fe, $Al_{12}(Fe,Mn)_3Si$, Al_2CuMg, Al_2Cu
Al−Cu−Li	Al_7Cu_2Fe, Al_2CuLi, Al_6CuLi_3

to form dispersoids due to their extremely low solubility in aluminum. Dispersoids are generally fewer than 1 μm in size and are uniformly distributed in matrix due to the slow diffusivity of Mn, Zr, and Cr in aluminum. Fine distribution of dispersoids controls recrystallization in aluminum alloys by limiting the motion of grain boundaries due to their grain boundary pinning ability. $Al_{20}Cu_2Mn_3$ in 2X24 alloys and Al_3Zr in Al−Li alloys are the main dispersoids.

A list of commonly observed constituent phases and dispersoids in 2XXX alloys is given in Table 2.4.

2.4.5 Effect of Predeformation

In precipitation-strengthened aluminum alloys, the density (distribution) and size of strengthening precipitates can significantly affect the strength of alloy. Aging practices such as soaking time and temperature control the size and density of precipitates. However, in some cases, plastic deformation during the interval between quenched state from solution treatment and artificial aging (collectively known as T8 temper) can also influence the precipitation kinetics. The influence of T8 temper path on precipitation hardening response depends upon the alloy system. 6XXX and 7XXX alloys are strengthened by precipitation of β' (Mg$_2$Si) and η' (MgZn$_2$) phases. Both, β' and η' precipitates, are considered coherent with the aluminum matrix and nucleate easily and homogenously in the matrix. So, introduction of dislocations prior to artificial aging in 6XXX or 7XXX alloys has little to no impact on hardening. As a matter of fact, Deschamps et al. [47] demonstrated that high amount (6–10%) of predeformation in an Al–Zn–Mg alloy can reduce the peak hardness. This response was attributed to the precipitation of coarse and less potent strengthener, η (equilibrium phase of η') on dislocations which eventually reduced the solute availability for primary strengthening precipitate η'.

In case of Al–Cu, Al–Cu–Mg, and Al–Cu–Li (2XXX) alloys, the main strengthening phases, θ' (Al$_2$Cu), S' or S (Al$_2$CuMg), and T$_1$ (Al$_2$CuLi), are considered semicoherent. These precipitates nucleate rather easily on heterogeneous sites such as grain and subgrain boundaries and dislocations. Thus, introduction of predeformation to these alloys results in enhanced hardening response upon artificial aging. Fig. 2.11 shows the effect of predeformation on aging response in various Al–Cu-based alloys [39,48].

Almost all the 2XXX alloys are known to benefit from predeformation, however, Ag containing Al–Cu–Mg alloys with high Cu:Mg ratio are an exception. Ringer et al. [48] studied the effect of cold work on various Al–Cu-based alloys and found that Ag containing

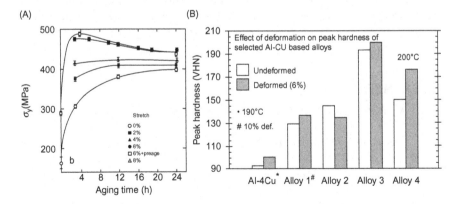

Figure 2.11 (A) Effect of plastic deformation on an Al–Cu–Li alloy [39] and (B) the effect of cold work prior to aging on peak hardness on selected Al–Cu-based alloys aged at 200°C: alloy 1, Al–4Cu–0.3Mg; alloy 2, Al–4Cu–0.3Mg–0.4Ag; alloy 3, Al–5.3Cu–1.3Li–0.4Mg–0.4Ag–0.12Zr; alloy 4, Al–3Cu–2Li–0.12Zr [48].
Source: Reprinted with permission from Springer.

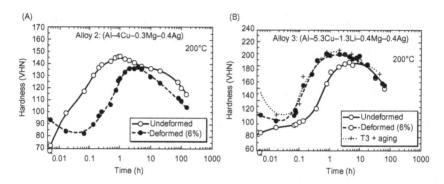

Figure 2.12 Effect of predeformation on hardness evolution during aging in (A) Al−Cu−Mg−Ag and (B) Al−Cu−Li−Mg−Ag alloys [48]. Source: Reprinted with permission from Springer.

predeformed Al−Cu−Mg alloys show delayed aging kinetics and lower peak hardness as compared to undeformed alloy (Fig. 2.12). As discussed earlier in Section 1.2.2, high Cu:Mg ratio Al−Cu−Mg alloys with Ag addition result in the formation of Ω precipitate. Ringer et al. [48] demonstrated that delayed aging response of Al−Cu−Mg−Ag alloy was primarily due to interference of dislocations with nucleation of Ω precipitates and promotion of nucleation of less potent but competing phase θ'. They found that cold working prior to aging increased the density of θ' at the expense of Ω precipitate. Also, dislocations are believed to disrupt the clusters of atoms which are essential for nucleation of Ω precipitate.

As described earlier, 2XXX aluminum alloys can be subcategorized into several different alloy systems due to different precipitates with a wide range of properties. Hence, friction stir welding (FSW) of 2XXX alloys will be discussed in two separate chapters: FSW of Al−Cu and Al−Cu−Mg alloys and FSW of Al−Cu−Li alloys.

REFERENCES

[1] J.D. Minford, Handbook of Aluminum Bonding Technology and Data, CRC Press, USA, 1993.

[2] G.E. Totten, D.S. MacKenzie, Handbook of Aluminum: Vol. 1: Physical Metallurgy and Processes, CRC Press, USA, 2003.

[3] E. Starke, J. Staley, Application of modern aluminum alloys to aircraft, Prog. Aerospace Sci 32 (1996) 131−172.

[4] T. Dursun, C. Soutis, Recent developments in advanced aircraft aluminium alloys, Mater. Des 56 (2014) 862−871.

[5] I. Fridlyander, Aluminum alloys with lithium and magnesium, Metal Sci. Heat Treat. 45 (2003) 344−347.

[6] Aluminum Association, Aluminum Standards and Data, Aluminum Association, 2000.

[7] R.N. Lumley, 1—Introduction to aluminium metallurgy, in: R. Lumley (Ed.), Fundamentals of Aluminium Metallurgy, Woodhead Publishing, UK, 2011, pp. 1–19.

[8] N.E. Prasad, A. Gokhale, R. Wanhill, Aluminum–Lithium Alloys: Processing, Properties, and Applications, Butterworth-Heinemann, UK, 2013.

[9] K. Sankaran, N. Grant, The structure and properties of splat-quenched aluminum alloy 2024 containing lithium additions, Mater. Sci. Eng. 44 (1980) 213–227.

[10] E.J. Lavernia, N.J. Grant, Aluminium–lithium alloys, J. Mater. Sci. 22 (1987) 1521–1529.

[11] I. Fridlyander, A. Dobromyslov, E. Tkachenko, O. Senatorova, Advanced high-strength aluminum-base materials, Metal Sci. Heat Treat 47 (2005) 269–275.

[12] I. Fridlyander, L. Khokhlatova, N. Kolobnev, K. Rendiks, G. Tempus, Thermally stable aluminum–lithium alloy 1424 for application in welded fuselage, Metal Sci. Heat Treat 44 (2002) 3–8.

[13] I. Fridlyander, V. Sister, O. Grushko, V. Berstenev, L. Sheveleva, L. Ivanova, Aluminum alloys: promising materials in the automotive industry, Metal Sci. Heat Treat 44 (2002) 365–370.

[14] N Facts Super Lightweight External Tank, National Aeronautics and Space Administration/Marshall Space Flight Center Huntsville, Alabama, 35812 (2005).

[15] J.C. Williams, E.A. Starke, Progress in structural materials for aerospace systems, Acta Mater 51 (2003) 5775–5799.

[16] R. Xiao, X. Zhang, Problems and issues in laser beam welding of aluminum–lithium alloys, J. Manuf. Process 16 (2014) 166–175.

[17] R.S. Mishra, P.S. De, N. Kumar, Friction Stir Welding and Processing: Science and Engineering, Springer, USA, 2014.

[18] R.S. Mishra, M.W. Mahoney, Friction Stir Welding and Processing, ASM International, 2007.

[19] R.S. Mishra, Z. Ma, Friction stir welding and processing, Mater. Sci. Eng. R Reports 50 (2005) 1–78.

[20] A. Kostrivas, J. Lippold, Weldability of Li-bearing aluminium alloys, Int. Mater. Rev 44 (1999) 217–237.

[21] The Welding Institute, http://www.twi-global.com/technical-knowledge/job-knowledge/weldability-of-materials-aluminium-alloys-021/.

[22] S.P. Ringer, K. Hono, T. Sakurai, I.J. Polmear, Cluster hardening in an aged Al–Cu–Mg alloy, Scr. Mater. 36 (1997) 517–521.

[23] S. Ringer, T. Sakurai, I. Polmear, Origins of hardening in aged Al–Cu–Mg–(Ag) alloys, Acta Mater 45 (1997) 3731–3744.

[24] S. Wang, M. Starink, Precipitates and intermetallic phases in precipitation hardening Al–Cu–Mg–(Li) based alloys, Int. Mater. Rev. 50 (2005) 193–215.

[25] G. Sha, R.K.W. Marceau, S.P. Ringer, 12—Precipitation and solute clustering in aluminium: advanced characterisation techniques, in: R. Lumley (Ed.), Fundamentals of Aluminium Metallurgy, Woodhead Publishing, UK, 2011, pp. 345–366.

[26] T. Gladman, Precipitation hardening in metals, Mater. Sci. Technol. 15 (1999) 30–36.

[27] J. Silcock, The effect of quenching on the formation of GP zones and θ' in Al Cu alloys, Philos. Mag 4 (1959) 1187–1194.

[28] S. Ringer, K. Hono, Microstructural evolution and age hardening in aluminium alloys: atom probe field-ion microscopy and transmission electron microscopy studies, Mater. Charact 44 (2000) 101–131.

[29] P. Villars, A. Prince, H. Okamoto, Handbook of ternary alloy phase diagrams, ASM Int. (1995).

[30] R.K.W. Marceau, G. Sha, R. Ferragut, A. Dupasquier, S.P. Ringer, Solute clustering in Al–Cu–Mg alloys during the early stages of elevated temperature ageing, Acta Mater 58 (2010) 4923–4939.

[31] R.K.W. Marceau, C. Qiu, S.P. Ringer, C.R. Hutchinson, A study of the composition dependence of the rapid hardening phenomenon in Al–Cu–Mg alloys using diffusion couples, Mater. Sci. Eng. A 546 (2012) 153–161.

[32] G. Sha, R.K.W. Marceau, X. Gao, B.C. Muddle, S.P. Ringer, Nanostructure of aluminium alloy 2024: segregation, clustering and precipitation processes, Acta Mater. 59 (2011) 1659–1670.

[33] K. Knowles, W. Stobbs, The structure of {1 1 1} age-hardening precipitates in Al–Cu–Mg–Ag alloys, Acta Crystallograph. B Struct. Sci. 44 (1988) 207–227.

[34] S. Ringer, W. Yeung, B. Muddle, I. Polmear, Precipitate stability in Al–Cu–Mg–Ag alloys aged at high temperatures, Acta Metal. Mater. 42 (1994) 1715–1725.

[35] B. Noble, G. Thompson, Precipitation characteristics of aluminium–lithium alloys, Metal Sci. J. 5 (1971) 114–120.

[36] B. Noble, G. Thompson, T 1 (Al$_2$CuLi) precipitation in aluminium–copper–lithium alloys, Metal Sci. J. 6 (1972) 167–174.

[37] D. Williams, J. Edington, The precipitation of δ'(Al$_3$Li) in dilute aluminium–lithium alloys, Metal Sci 9 (1975) 529–532.

[38] A. Deschamps, C. Sigli, T. Mourey, F. De Geuser, W. Lefebvre, B. Davo, Experimental and modelling assessment of precipitation kinetics in an Al–Li–Mg alloy, Acta Mater 60 (2012) 1917–1928.

[39] W. Cassada, G. Shiflet, E. Starke, The effect of plastic deformation on Al$_2$CuLi (T 1) precipitation, Metal. Trans. A 22 (1991) 299–306.

[40] A.K. Shukla, W.A. Baeslack III, Study of microstructural evolution in friction-stir welded thin-sheet Al–Cu–Li alloy using transmission-electron microscopy, Scr. Mater. 56 (2007) 513–516.

[41] A.K. Shukla, Friction stir welding of thin-sheet, age-hardenable aluminum alloys: a study of process/structure/property relationships, ProQuest Dissertations and Theses (2007).

[42] A. Shukla, W. Baeslack III, Study of process/structure/property relationships in friction stir welded thin sheet Al–Cu–Li alloy, Sci. Technol. Weld. Join. 14 (2009) 376–387.

[43] H.H. Jo, K. Hirano, Precipitation Processes in Al–Cu–Li Alloy Studied by DSC, 13 (1987) 377–382.

[44] B. Decreus, A. Deschamps, F. De Geuser, P. Donnadieu, C. Sigli, M. Weyland, The influence of Cu/Li ratio on precipitation in Al–Cu–Li–X alloys, Acta Mater 61 (2013) 2207–2218.

[45] G.H. Narayanan, W. Quist, B. Wilson, A. Wingert, Low density aluminum alloy development, AFWAL Contract F33615-81-c-5053 (1982).

[46] B. Huang, Z. Zheng, Independent and combined roles of trace Mg and Ag additions in properties precipitation process and precipitation kinetics of Al–Cu–Li–(Mg)–(Ag)–Zr–Ti alloys, Acta Mater 46 (1998) 4381–4393.

[47] A. Deschamps, F. Livet, Y. Brechet, Influence of predeformation on ageing in an Al–Zn–Mg alloy—I. Microstructure evolution and mechanical properties, Acta Mater 47 (1998) 281–292.

[48] S. Ringer, B. Muddle, I. Polmear, Effects of cold work on precipitation in Al–Cu–Mg–(Ag) and Al–Cu–Li–(Mg–Ag) alloys, Metal. Mater. Trans. A 26 (1995) 1659–1671.

CHAPTER 3

Temperature Evolution and Thermal Management During FSW of 2XXX Alloys

3.1 TEMPERATURE EVOLUTION

In case of aluminum alloys, FSW is a thermomechanical process occurring at high temperatures, usually highest temperature lying in the range of $0.6T_m - 0.9T_m$ (T_m = melting temperature of alloy) [1–3]. 2XXX aluminum alloys are precipitation-strengthened alloy. So, depending on time and temperature of thermal treatment, 2XXX alloys exhibit a wide range of mechanical properties due to combination of different type, distribution, and morphology of precipitates. Thus, it is appropriate to acknowledge the temperature field evolving in various zones during FSW as a key factor influencing the post-weld properties.

During FSW, total heat generated is a combination of frictional heat and adiabatic heat [1,2]. Frictional heat is generated due to the friction between tool material and workpiece material, where a major fraction of heat comes from tool shoulder interaction with workpiece. Another contribution to heat is due to the plastic deformation of metal. A part of the plastic deformation energy is stored in the thermo-mechanically affected zone in the form of increased dislocation (defects) density. The temperature generation during FSW is influenced by welding parameters, geometry, and dimensions of tool shoulder, tool pin, workpiece thickness, and thermal conductivities of workpiece and anvil (welding bed or support structure). Generally, welding parameters can be empirically correlated to temperature; high tool rotation rate and low welding speed produce hotter welds, whereas low tool rotation rate and high welding speed result in colder welds.

In available literature, temperature evolution during FSW of various aluminum alloys have been either experimentally captured using thermocouples embedded into the workpiece (or sometimes into the tool) or calculated using various numerical modeling techniques. Both

Friction Stir Welding of 2XXX Aluminum Alloys Including Al–Li Alloys.
DOI: http://dx.doi.org/10.1016/B978-0-12-805368-3.00003-0

the techniques have their own merits and limitations. While embedded thermocouple measurements are the most accurate way of capturing the thermal profile of a specific location, the exact temperature from the weld nugget cannot be captured easily. Numerical modeling provides an advantage with thermal history of all the locations simulated during modeling, but often deviate from actual thermal history. Experimentally captured or simulated peak temperature, reported in literature, in various zones (WN, TMAZ, and HAZ) during FSW of various 2XXX alloys are summarized in Table 3.1.

Table 3.1 Peak Temperature Reported in Literature in Various Zones of FSW of 2XXX Alloys

Peak Temperature (°C)			RPM-IPM	S_D, W_D (mm)	Workpiece Thickness (mm)	Alloy	Reference
WN	TMAZ	HAZ					
		350	400-4		3.2	2024-T3	[4]
		330	400-16		3.2	2024-T3	[4]
	335		400-3.2	8, 1.6	1.6	2024-T3	[5]
	400		700-8	8, 1.6	1.6	2024-T3	[5]
	435		1500-40	8, 1.6	1.6	2024-T3	[5]
	455		1200-48	8, 1.6	1.6	2024-T3	[5]
		330	650-2.4	19, 5.5	6.5	2024	[6]
		140	650-2.4	19, 5.5	6.5	2024	[6]
		340	850-4.8	16, 5.5	6	2024-T351	[7]
325[a]			215-3		6.4	2024-T351	[8]
480[a]			360-8		6.4	2024-T351	[8]
515[a]			360-10		6.4	2024-T351	[8]
305[a]			120-2		6.4	2524-T351	[8]
380[a]			300-5		6.4	2524-T351	[8]
400[a]			480-8		6.4	2524-T351	[8]
452			1547-1		4	2024-T6	[9]
433			1547-1		4	2024-O	[9]
	481		468-3	20, 6.35	6.35	2024-T351	[10]
		250	215-6	20, 6.35	6.35	2024-T351	[10]
		275	468-6	20, 6.35	6.35	2024-T351	[10]
		300	350-4	20, 6.35	6.35	2024-T351	[10]
		272	215-3	20, 6.35	6.35	2024-T351	[10]
		330	468-3	20, 6.35	6.35	2024-T351	[10]
		252	400-1.7	11[b], 3.2	3.2	2198-T851	[11]

(Continued)

Table 3.1 (Continued)

Peak Temperature (°C)			RPM-IPM	S_D, W_D (mm)	Workpiece Thickness (mm)	Alloy	Reference
WN	TMAZ	HAZ					
		267	600-1.7	11[b], 3.2	3.2	2198-T851	[11]
		302	800-1.7	11[b], 3.2	3.2	2198-T851	[11]
		330	1000-1.7	11[b], 3.2	3.2	2198-T851	[11]
		340	800-16	20, 4.8	6.5	2024-T351	[12]
		340	800-4	20, 4.8	6.5	2024-T351	[12]
		217	800-16	15, 3.05	3.2	2024-T351	[13]
450			800-16	15, 3.05	3.2	2024-T351	[13]
460			1200-16	15, 3.05	3.2	2024-T351	[13]
467			1600-16	15, 3.05	3.2	2024-T351	[13]
467	420	308	800-8	14, 4.8	5	2219-T6	[14]
380[a]			1800-12	7, 0.9	1	2195-T8	[15]
422[a]			2400-3	7, 0.9	1	2195-T8	[15]
323[a]			1800-18	7, 0.9	1	2024-T3	[16]
428[a]			2600-2	7, 0.9	1	2024-T3	[16]
	392	315	550-3.2	28, 13.6	14	2219-O	[17]
	401	323	600-3.2	28, 13.6	14	2219-O	[17]
	393	309	600-4	28, 13.6	14	2219-O	[17]
	420		800-4	22.5, 7.4	7.5	2219-T6	[18]
	398[c]		800-4	22.5, 7.4	7.5	2219-T6	[18]

WN, weld nugget; TMAZ, thermomechanically affected zone; HAZ, heat affected zone; RPM, revolutions per minute; IPM, inches per minute; SD, tool shoulder diameter; WD, weld depth
[a]*Numerically simulated values*
[b]*Bobbin tool*
[c]*Welding submerged in water*

Thermal conductivities among 2XXX aluminum alloys are quite similar. Therefore, under similar conditions (tool, welding parameters, material thickness), temperature evolution during FSW of various 2XXX alloys will be similar. Peak temperature in weld nugget ranges from 350°C to 540°C (Table 3.1). The highest possible peak temperature in weld nugget during FSW of any 2XXX alloy remains around the solidus temperature of the alloy. In FSW, temperature rises due to frictional heat and energy dissipated due to the resistance of material to plastic deformation. As the temperature rises close to solidus temperature, the shear strength of the alloy reduces significantly, and incipient melting may occur. This diminishes both frictional and plastic deformation heat generation during FSW and hence further temperature rise

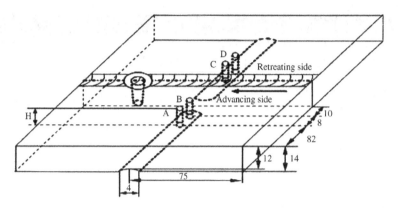

Figure 3.1 Schematic of welding setup used by Xu et al. [17] to study the temperature evolving at various locations, during FSW of a thick 2219 alloy. Source: Reprinted with permission from Elsevier.

will not occur. Solidus temperature of an alloy depends upon the alloying content, and it varies from 500°C to 560°C in 2XXX aluminum alloys [19,20].

Xu et al. [17] comprehensively studied the temperature evolution in 14-mm thick aluminum alloy 2219-O during FSW. They used K-type thermocouple embedded into workpiece to measure the thermal profiles as a function of depth of the location, tool rotation rate, and tool traverse speed. Fig. 3.1 shows the details of locations of thermocouples used to capture the thermal profile during FSW. Location H denotes the distance between a point and bottom of the weld along the depth of the weld. Locations A and D correspond to HAZ on advancing and retreating sides, while B and C are for TMAZ on advancing and retreating sides.

Fig. 3.2 shows the temperature plots of different locations of point B (as shown in Fig. 3.1) along the thickness of workpiece during FSW. Note that point $H = 12$ mm is close to the top of the weld, and $H = 2$ mm is close to the bottom of the weld. Fig. 3.2 clearly shows that the peak temperature decreases along the depth of the weld. The peak temperature gradient along the thickness of workpiece is more than 20°C. This is due to the fact that the shoulder provides major contribution to the heat generation. Also, total surface area of the pin decreases along the weld, which reduces the size of deformation zone along the depth, thereby reducing the heat content along the depth. Another key observation from Fig. 3.2 is that the heating and cooling rates also decreases significantly along the depth of the weld. Such gradients in peak

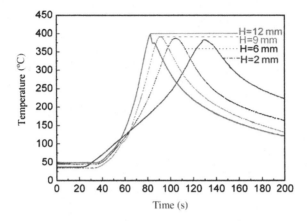

Figure 3.2 Temperature evolution variations as a function of depth along the depth of workpiece during FSW of thick 2219 alloy [17]. Source: Reprinted with permission from Elsevier.

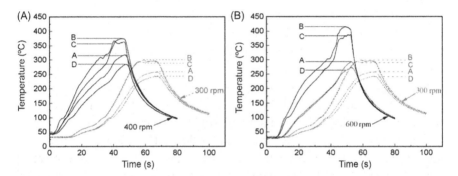

Figure 3.3 Temperature curves for different tool rotation speeds during tool inserting: (A) 300 and 400 rpm; (B) 300 and 600 rpm [17]. Source: Reprinted with permission from Elsevier.

temperature, heating rates, and cooling rates increase with increase in thickness of the workpiece (or weld). In case of thin sheets, the gradient is negligible as the separation between tool shoulder and bottom of the weld is very small.

The thermal profile is greatly influenced by the choice of FSW parameters (tool rotation rate and traverse speed). As discussed earlier, peak temperature of any location during welding increases with increase in tool rotation rate. This is attributed to the fact that the frictional heat increases due to higher tool rotation rate. Fig. 3.3 shows the temperature profiles from the study by Xu et al. [17], comparing effect of tool rotation rate on temperature evolution at various locations. While increase in welding speed (traverse speed) can reduce the

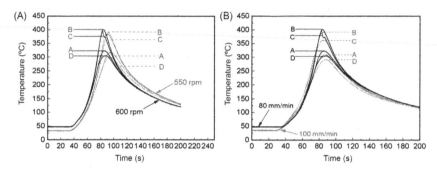

Figure 3.4 Temperature profiles for different parameters during stable welding: (A) different tool rotation speeds (traverse speed: 80 mm/min) and (B) different traverse speeds (rotary speed: 600 rpm) [17]. Source: Reprinted with permission from Elsevier.

peak temperature, the extent of reduction in temperature is usually low. The temperature profiles of various locations as function of different tool rotation rate (Figs 3.3 and 3.4a) and traverse speed (Fig. 3.4b). Clearly, change in tool rotation rate has a larger impact of peak temperature as compared to change in traverse speed. Increasing traverse speed increases the heating and cooling rates. Overall, the increase in ratio of tool rotation speed to traverse speed increases the peak temperature during FSW of 2XXX alloys. The temperature distribution in workpiece is considered slightly asymmetric, with slightly higher temperatures on the retreating side of the weld [2]. However, the difference between peak temperature in advancing side and retreating side is small, and can be reduced to a great extent with improved tool design.

3.2 THERMAL MANAGEMENT

Precipitation-strengthened aluminum alloys are known to suffer a knockdown in strength after FSW. Thermal cycles experienced during welding significantly impact the extent of knockdown. It is clear from the experimental data presented in Table 3.1 that the peak temperature in all the three zones of weld are high enough for phase transformations to occur in 2XXX alloys. As the field of FSW advances, various thermal management techniques are being explored to influence the post-weld microstructure and properties to manipulate the thermal cycle [21–25]. Fig. 3.5 shows a schematic representation of various thermal management techniques which can be utilized to influence the welding process and postweld properties.

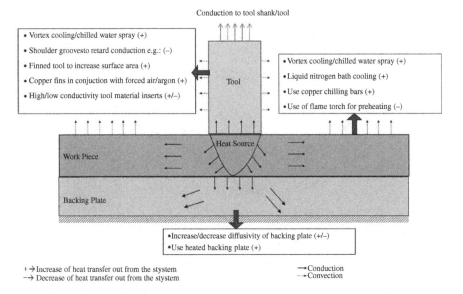

Figure 3.5 Schematic showing thermal management methods that can be used in friction stir welding process. Arrows indicate heat transfer [23]. Source: Reprinted with permission from Dr. Piyush Upadhyay.

From various methods shown in Fig. 3.5, use of high thermal conductivity backing plate is the most utilized method. A backing plate of copper or aluminum can be used to manipulate the thermal cycle to enhance the post-weld mechanical properties [22]. The effectiveness of use of any thermal management method is highly dependent on the material to be joined. For example, Upadhyay [23] in his dissertation showed that through thickness microstructural homogeneity can be attained by using low thermal conductivity backing plate like ceramic during FSW of a 4.2-mm thick AA6056-T4 alloy. In another work [22], the authors of this book have also showed the effectiveness of using a copper backing with high thermal conductivity during FSW of 1424, an Al-Mg-Li alloy. However, using a high thermal conductivity backing plate may also result in defective welds due to excessive heat reduction close to bottom of the weld, as shown by Zhang et al. [26]. Passive heating or cooling of workpiece have also been demonstrated and can be optimized to affect post-weld properties [27].

Zhang et al. [25] demonstrated the use of an arrangement of backing plates to be more efficient as compared to a single backing with high thermal conductivity. They used a combination of copper and medium carbon steel as backing plate to target different zones of FSW. The schematic of the arrangement used by Zhang et al. [25] is

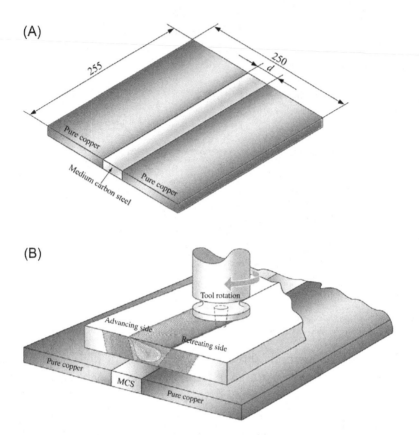

Figure 3.6 (A) and (B) Schematic showing arrangement of composite backing plate used by Zhang et al. [25] to manipulate thermal cycle during FSW. Source: Reprinted with permission from Elsevier.

shown in Fig. 3.6. While high conductivity copper was used to enhance the cooling rates in TMAZ and HAZ, where an insert of medium carbon steel between copper plates was used to sustain high temperature in weld nugget for efficient solutionization.

REFERENCES

[1] R.S. Mishra, Z. Ma, Friction stir welding and processing, Mater. Sci. Eng. Rep. 50 (2005) 1–78.

[2] R.S. Mishra, M.W. Mahoney, Friction Stir Welding and Processing, ASM International, 2007.

[3] R.S. Mishra, P.S. De, N. Kumar, Friction Stir Welding and Processing: Science and Engineering, Springer, 2014.

[4] E. Hersent, J.H. Driver, D. Piot, C. Desrayaud, Integrated modelling of precipitation during friction stir welding of 2024-T3 aluminium alloy, Mater. Sci. Technol. 26 (2010) 1345–1352.

[5] R. Fu, J. Zhang, Y. Li, J. Kang, H. Liu, F. Zhang, Effect of welding heat input and post-welding natural aging on hardness of stir zone for friction stir-welded 2024-T3 aluminum alloy thin-sheet, Mater. Sci. Eng. A 559 (2013) 319–324.

[6] S. Benavides, Y. Li, L. Murr, D. Brown, J. McClure, Low-temperature friction-stir welding of 2024 aluminum, Scr.Mater. 41 (1999) 809–815.

[7] C. Genevois, D. Fabrègue, A. Deschamps, W.J. Poole, On the coupling between precipitation and plastic deformation in relation with friction stir welding of AA2024 T3 aluminium alloy, Mater. Sci. Eng. A 441 (2006) 39–48.

[8] B. Yang, J. Yan, M.A. Sutton, A.P. Reynolds, Banded microstructure in AA2024-T351 and AA2524-T351 aluminum friction stir welds: Part I. Metallurgical studies, Mater. Sci. Eng. A 364 (2004) 55–65.

[9] M. Mohammadtaheri, M. Haddad-Sabzevar, M. Mazinani, E.B. Motlagh, The effect of base metal conditions on the final microstructure and hardness of 2024 aluminum alloy friction-stir welds, Metall. Mater. Trans. B 44 (2013) 738–743.

[10] M. Jariyaboon, A. Davenport, R. Ambat, B. Connolly, S. Williams, D. Price, The effect of welding parameters on the corrosion behaviour of friction stir welded AA2024–T351, Corros. Sci. 49 (2007) 877–909.

[11] F. Wang, W. Li, J. Shen, S. Hu, J. dos Santos, Effect of tool rotational speed on the microstructure and mechanical properties of bobbin tool friction stir welding of Al–Li alloy, Mater. Des 86 (2015) 933–940.

[12] Z. Zhang, B. Xiao, Z. Ma, Hardness recovery mechanism in the heat-affected zone during long-term natural aging and its influence on the mechanical properties and fracture behavior of friction stir welded 2024Al–T351 joints, Acta Mater 73 (2014) 227–239.

[13] P.A. Colegrove, H.R. Shercliff, R. Zettler, Model for predicting heat generation and temperature in friction stir welding from the material properties, Sci. Technol. Weld. Joining 12 (2007) 284–297.

[14] Y. Chen, J. Feng, H. Liu, Precipitate evolution in friction stir welding of 2219-T6 aluminum alloys, Mater. Charact 60 (2009) 476–481.

[15] A. Shukla, W. Baeslack III, Study of process/structure/property relationships in friction stir welded thin sheet Al–Cu–Li alloy, Sci. Technol. Weld. Joining 14 (2009) 376–387.

[16] AK Shukla, Friction stir welding of thin-sheet, age-hardenable aluminum alloys: a study of process/structure/property relationships, ProQuest Dissertations and Theses 2007.

[17] W. Xu, J. Liu, G. Luan, C. Dong, Temperature evolution, microstructure and mechanical properties of friction stir welded thick 2219-O aluminum alloy joints, Mater. Des 30 (2009) 1886–1893.

[18] H. Zhang, H. Liu, L. Yu, Effect of water cooling on the performances of friction stir welding heat-affected zone, J. Mater. Eng. Perform 21 (2012) 1182–1187.

[19] JR Davis, ASM Specialty Handbook: Aluminum and Aluminum Alloys, 1993.

[20] G.E. Totten, D.S. MacKenzie, Handbook of Aluminum: Vol.1: Physical Metallurgy and Processes, CRC Press, 2003.

[21] P. Upadhyay, A. Reynolds, Effect of backing plate thermal property on friction stir welding of 25-mm-thick AA6061, Metall. Mater. Trans. A 45 (2014) 2091–2100.

[22] H. Sidhar, N.Y. Martinez, R.S. Mishra, J. Silvanus, Friction stir welding of Al–Mg–Li 1424 alloy, Mater. Des 106 (2016) 146–152.

[23] P. Upadhyay, Thermal management in friction stir welding of aluminum alloys. (2012).

[24] P. Upadhyay, A.P. Reynolds, Thermal management in friction-stir welding of precipitation-hardened aluminum alloys, JOM 67 (2015) 1022–1031.

[25] Z. Zhang, W. Li, Y. Feng, J. Li, Y. Chao, Improving mechanical properties of friction stir welded AA2024-T3 joints by using a composite backplate, Mater. Sci. Eng. A 598 (2014) 312–318.

[26] Z. Zhang, W. Li, J. Shen, Y. Chao, J. Li, Y. Ma, Effect of backplate diffusivity on microstructure and mechanical properties of friction stir welded joints, Mater. Des 50 (2013) 551–557.

[27] T. Nelson, R. Steel, W. Arbegast, In situ thermal studies and post-weld mechanical properties of friction stir welds in age hardenable aluminium alloys, Sci. Technol. Weld. Joining (2013).

CHAPTER 4

FSW of Al–Cu and Al–Cu–Mg Alloys

4.1 INTRODUCTION

Friction stir welding (FSW) of precipitation-strengthened aluminum alloys results in a variety of microstructure and mechanical properties. Of course, initial microstructure and alloy chemistry plays the most important role in determining the final microstructure. Factors such as choice of welding parameters, thermal cycle during welding, use of auxiliary cooling systems, tool geometry, and workpiece thickness also greatly influence the resultant microstructure and strength of the weld.

As discussed in earlier chapters, 2XXX aluminum alloys consist of a wide range of precipitates. Precipitates in 2XXX alloys vary in stability, morphology, coherency, structure, and strengthening potential depending on the alloy chemistry. For example, GP zones of Al–Cu–Mg system are considered more stable as compared to GP zones in Al–Cu system. Similarly, T_1 (Al_2CuLi) precipitate provides the highest strength in 2XXX alloys whereas δ' (Al_3Li) phase is least potent strengthener. Therefore, the type of precipitates in an alloy affect the postweld microstructure the most. Differential scanning calorimetry (DSC) has been extensively used to study the precipitation kinetics in various precipitation-strengthened alloys [1–3]. Table 4.1 shows the approximate transformation temperature ranges for various precipitates of 2XXX alloys. Note that the data provided is solely based on constant temperature ramp DSC experiment results. Therefore, transformation temperature during other thermomechanical treatments will vary. It is clear from Table 4.1 that during a typical thermal cycle of FSW, postweld microstructure will be affected by the type of precipitates in the workpiece material. Hence, majority of the discussion on FSW of 2XXX alloy will be focused on the precipitates and their transformation.

Friction Stir Welding of 2XXX Aluminum Alloys Including Al–Li Alloys.
DOI: http://dx.doi.org/10.1016/B978-0-12-805368-3.00004-2

Table 4.1 Precipitation and Dissolution Temperature Range of Various Precipitates in 2XXX Alloys

Phases	Temperature Range (°C)	
	Precipitation [References]	Dissolution [References]
GP zones (Al−Li)	RT−100 [1]	75 and above [1,2]
GP zones (Al−Cu)	RT−150 [4,5]	100 and above [4,5]
GPB zones	RT−160 [6]	180 and above [6]
δ' (Al$_3$Li)	RT−200 [1,7]	180 and above [1]
θ' (Al$_2$Cu)	200−320 [4]	320 and above [4]
S (Al$_2$CuMg)	260−320 [6]	320 and above [6]
T$_1$ (Al$_2$CuLi)	250−350 [8]	350 and above [8]

Note that the temperature range data is based on DSC results, therefore transformation temperature during isothermal heat treatments may vary.

4.2 FSW OF Al−Cu ALLOYS

Al−Cu alloys are primarily strengthened by GP zones in naturally or underaged (T3 or T4) condition, and θ' precipitates provide strength in peak aged conditions (T6 or T8). Alloy 2219 is the most commonly used alloy of this subcategory. A number of studies on various aspects of FSW of Al−Cu alloys exist [9−23]. Alloy 2219 in various tempers has been used in these studies.

4.2.1 Microstructural Evolution

The microstructural response of any precipitation-strengthened aluminum alloy is highly influenced by the starting temper (initial microstructure) of the material. Also, FSW results in three microstructural zones as explained in detail in earlier chapters. Hence, microstructural evolution in different zones will be dealt separately.

4.2.1.1 Weld Nugget

During FSW, weld nugget (WN) goes through extreme deformation and thermal cycles. The strain in WN usually ranges from 2 to 5 [25]. The temperature in WN usually reaches the solution temperature range of aluminum alloys [24,25]. Chen et al. [19] studied the precipitate evolution in friction stir welded 2219-T6 alloy. Temperature in WN usually lies in the range of 400−550°C as demonstrated by Chen et al. for alloy 2219 [19]. Such temperatures are high enough to dissolve θ' precipitates present in parent material of T6 temper as clearly demonstrated in Fig. 4.1A which shows solvus boundaries of various precipitates in Al−Cu system. Fig. 4.1A also shows that solvus temperature

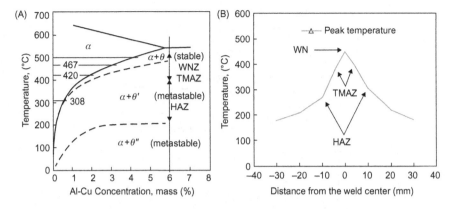

Figure 4.1 (A) Phase diagram of Al–Cu system showing solvus of various phases and peak temperatures in various zones of FSW [19] and (B) experimentally measured temperature at various locations during FSW [19]. Source: Reprinted with permission from Elsevier.

for formation of complete solid solution in Al–Cu alloy changes with Cu content. Alloy 2219 contains high amount of Cu (> 6 wt%). So, the phase diagram in Fig. 4.1A suggests that even high temperatures of 500–525°C are not enough to form complete solid solution in WN of high Cu containing Al–Cu alloys.

Overall, partial or full dissolution of θ' precipitates during thermal and deformation cycles of FSW produces partial to complete solid solution in WN depending on the peak temperature during welding. However, cooling rates during and after FSW also affect the final microstructure in WN in as-welded condition. If the cooling rates are low, reprecipitation of second phases can occur. Fig. 4.2 shows a time–temperature–transformation (TTT) diagram of an Al–4wt%Cu alloy [26]. It is clear from Fig. 4.2 that extremely high cooling rates in the range of 200–250°C are required to avoid the "nose" of θ phase formation. In conventional FSW process, such high cooling rates cannot be achieved and thus, reprecipitation of second phases during the cooling cycle in WN can also occur. Many researchers have reported similar findings in WN of 2219 alloy [3,19,27]. Fig. 4.3 shows transmission electron microscopy (TEM) images from WN of 2219 alloy reported in various studies [18,19]. Fig. 4.3 shows TEM images of WN of 2219 alloy in AW condition, which show largely precipitate free matrix with a few sparsely nucleated θ phase during cooling cycle of the welding process. Note that the reprecipitated phase is usually stable θ phase which is less potent strengthener as compared to

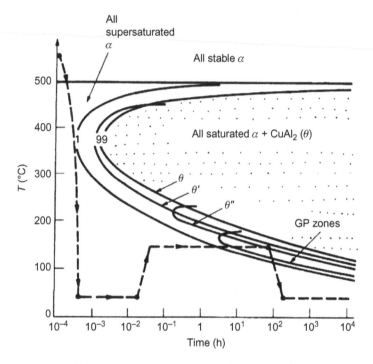

Figure 4.2 TTT diagram of Al–4wt%Cu system [26]. Source: Reprinted with permission from Elsevier.

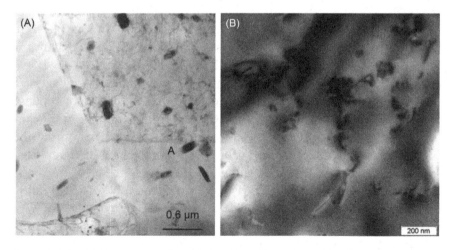

Figure 4.3 (A) and (B) TEM images of WN of FSWed 2219 alloy in AW condition showing sparesely distributed θ phase [18,19]. Source: (A) Reprinted with permission from Elsevier. (B) Reprinted with permission from Springer.

metastable θ' phase which provides strengthening in peak aged Al–Cu alloys. This is due to the fact that stable precipitates are favorable to form at higher temperatures as evident from the TTT diagram. FSW of Al–Cu alloys in other tempers (T3, T4, T8, and O) also results in similar microstructure in WN in AW condition as described earlier [11,18,28]. Basically, thermal and deformation cycles during FSW result in similar microstructure. Though, weaker tempers such as T3, T4, and O tempers are relatively easier to weld. Coarse grain boundary precipitates can also form if the cooling rates are very slow. Grain boundaries act as favorable nucleation site for precipitates as it reduces the interfacial energy of incoherent precipitates.

Artificial aging is the key step in achieving peak strength in precipitation-strengthened alloys. Similarly, postweld heat treatment (PWHT) is utilized based on specific application. Strength in WN can be recovered to various extent. The solutionized matrix produced during FSW retains solutes which can form precipitates during PWHT. Though, the volume fraction of precipitates during PWHT depends upon the amount of solute available which is governed by the peak temperature and cooling rates during FSW as discussed earlier. Chen et al. [19] also studied the microstructural evolution in FSW of 2219 alloy during PWHT. Fig. 4.4 shows low and high magnification TEM images of WN aged for 18 hours at 165°C (PWHT). Fig. 4.4A shows

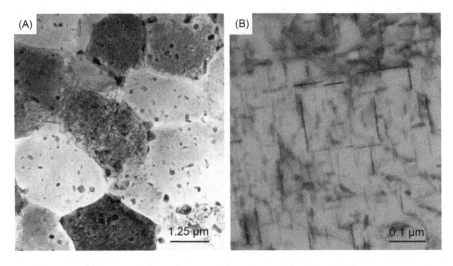

Figure 4.4 TEM image showing (A) θ precipitates and (B) θ' precipitates in the PWHT condition in WN of FSWed 2219-T6 alloy [19]. Source: Reprinted with permission from Elsevier.

the presence of coarse precipitates in the grain interior and boundaries. These phases were formed during cooling cycle and further precipitation and coarsening of these phases continued during PWHT. Fig. 4.4B shows the fine platelets of metastable θ' precipitate formed in WN during PWHT. It indicates that precipitation of both coarse stable θ and fine metastable θ' phases occurs during PWHT.

4.2.1.2 Thermomechanically Affected Zone

Thermomechanically affected zone (TMAZ) is the transition zone between WN and heat affected zone (HAZ). TMAZ may experience partial recrystallization in some cases if the temperature and strain are high enough. The grain morphology in TMAZ changes to highly deformed elongated grains. The temperature gradient across the transverse weld cross-section results in lower peak temperature in TMAZ as compared to WN. Chen et al. [19] experimentally measured peak temperature in TMAZ during FSW of 2219-T6 alloy. Peak temperature in TMAZ was observed in the range of 380–410°C as shown in Fig. 4.1B. According to Al–Cu phase diagram shown in Fig. 4.1A, dissolution of metastable θ' precipitates can occur if the material is in aged condition. However, the temperature range in TMAZ is not high enough for the formation of complete solid solution. Hence, as observed by Chen et al. [19], partial dissolution in TMAZ occurs and metastable precipitates are transformed into stable θ precipitates. Microstructures of TMAZ in AW and PWHT conditions are shown in Fig. 4.5A and B. These show presence of stable θ phase in TMAZ in AW condition. Reprecipitation of coarse precipitates also occurs during cooling cycle of the weld. After PWHT, these stable precipitates further grow as shown in Fig. 4.5C. Also, precipitation of thin plate-shaped θ' precipitates takes place during PWHT as shown in Fig. 4.5D [19]. It implies that considerable amount of solutes were retained in TMAZ after dissolution during FSW. After PWHT, the number density of reprecipitated θ' precipitates in the TMAZ (Fig. 4.5D) was less than that in the WN (Fig. 4.4B). It shows that the level of dissolution in WN compared to the TMAZ changes as a result of the difference in temperature between TMAZ and WN during FSW.

4.2.1.3 Heat Affected Zone

HAZ is the zone which only experiences thermal cycle. The temperature in HAZ can vary between 250°C and 350°C as experimentally measured by Chen et al. [19] (Fig. 4.1B). TTT diagram shown in

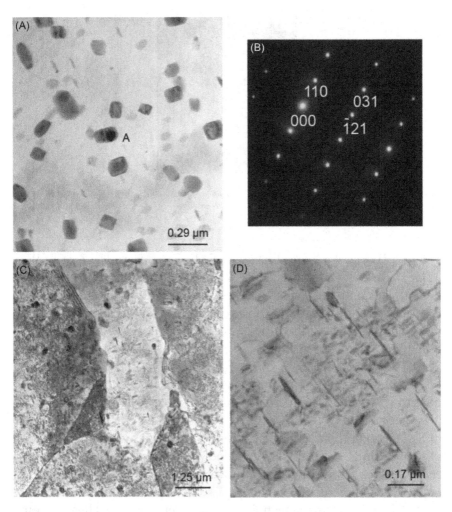

Figure 4.5 (A) TEM image of TMAZ in AW condition, (B) selected area diffraction pattern of precipitate marked "A" in (A), (C) low magnification TEM image showing θ precipitates and grains structure in the PWHT condition, and (D) high magnification TEM image showing the reprecipitated metastable precipitates in the PWHT condition of TMAZ of FSWed 2219 alloy [19]. Source: Reprinted with permission from Elsevier.

Fig. 4.2 shows that even short exposure time (usually a few seconds) at such high temperatures lies within the transformation range of θ' and θ phases. Moreover, the aging kinetics of θ' and θ phases are fastest in near 350°C (Fig. 4.2). This region of TTT diagram lies in the coarsening regime of precipitates. It implies that the region adjacent to TMAZ, which experiences temperature with enhanced kinetics of precipitates results in extensive coarsening of precipitates. Extensive coarsening is usually observed if the starting material is in aged condition

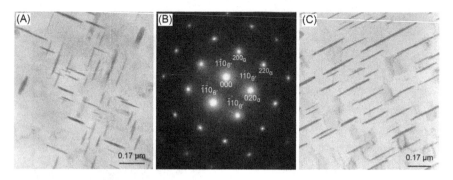

Figure 4.6 (A) TEM image of HAZ of FSWed 2219-T6 alloy in AW condition, (B) SADP of microstructure shown in (A), and (C) microstructure of HAZ in PWHT condition [19]. Source: Reprinted with permission from Elsevier.

(T6 or T8). Fig. 4.6A and B shows TEM images from HAZ of FSWed 2219-T6 in AW condition [19]. Bright field images clearly show a mixed distribution of plate-shaped precipitates. These precipitates coarsen due to thermal exposure during FSW. Chen et al. [19] reported that the plate-shaped precipitates in HAZ had coarsened about 20–40 nm in thickness as compared to 15–20 nm in BM of 2219-T6. Note that length of precipitates in HAZ also increases as a result of coarsening. Fig. 4.6B shows the selected area diffraction pattern (SADP) of the microstructure shown in Fig. 4.6A taken along [0 0 1] zone axis of aluminum. The presence of strong diffraction spots in SADP from θ' phase indicates the presence of a high volume fraction of θ' phase in HAZ region in AW condition. After PWHT, HAZ experiences further coarsening of precipitates. Microstructure of HAZ in PWHT condition is shown in Fig. 4.6C. Aging temperature during PWHT is usually high enough for growth of precipitates. Thus, on being subjected to PWHT, further coarsening of precipitates takes place.

In case of FSW of aged (T6, T7, or T8) material, extensive coarsening of precipitates in HAZ is usually inevitable, however, the extent of coarsening and width of HAZ can be manipulated to some degree with the use of auxiliary cooling systems. In case of FSW of underaged Al–Cu alloy (T3 or T4), GP zones provide strengthening as compared to θ' precipitates in aged material. Since temperature in HAZ during FSW is high enough for significant precipitation to occur in short duration as evident from TTT diagram shown in Fig. 4.2. Thus, coarsening of precipitates also occurs in HAZ of underaged material (T3 or T4). FSW of alloy 2219 in O temper has also been studied [20,28–30].

Material in O temper is as annealed and possesses lowest strength and highest ductility. Although detailed microstructure of HAZ of O temper material has not been studied, insignificant precipitation is anticipated from thermal cycle of FSW.

4.2.2 Mechanical Property Evolution
Strength of the weld of an Al—Cu alloy mainly depends on the initial microstructure and process parameters used during FSW. As discussed earlier, FSW results in microstructural variation in weld cross-section. It is understandable that microstructural variation results in mechanical property variation as well.

4.2.2.1 Hardness Evolution in AW Condition
Liu et al. [23] studied the effect of process parameters and PWHT on mechanical properties of FSWed 2219-T6 alloy. They performed FSW on 5-mm-thick 2219-T6 alloy using 800 rpm and welding speed was varied between 60 and 300 mm per minute (mmPM) [23]. Microhardness across the weld cross-section in AW condition was reported for the case of 140, 200, and 300 mmPM as shown in Fig. 4.7A. Hardness in WN is observed to be lower (108−113 HV) as compared to BM (122 HV) due to dissolution taking place during FSW [23]. In WN in AW condition, grain size strengthening and solid solution strengthening act as the key strengthening mechanisms. Natural aging can also impact the hardness with time. It is clear from Fig. 4.7A that lower welding speeds resulted in lower hardness in WN in AW condition. Hardness of WN was around 108 HV for welding speed of 140 mmPM, whereas welding

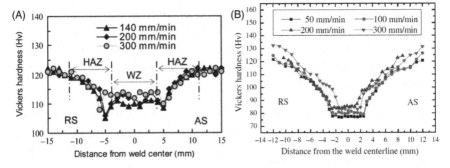

Figure 4.7 (A) and (B) Microhardness plots of FSWed 2219-T6 alloy made with tool rotation rate of 800 rpm, varying welding speed, and using tools with similar dimensions [13,23]. Note that the difference in hardness profile in both the cases is mainly due to use of tool with stationary shoulder in (B), which result in low heat input in HAZ as compared to conventional FSW tool with nonstationary shoulder. Source: (A) Reprinted with permission from Taylor & Francis. (B) Reprinted with permission from Springer.

speed of 300 mmPM resulted in hardness of 113 HV in WN [23]. It is due to the lower cooling rates associated with lower welding speed which result in formation of coarse second phases during cooling of the weld. Lower welding speed also produces coarser grain size which reduces the grain size strengthening contribution. It also reduces the quantity of solutes in solid solution thereby reducing the solid solution strengthening contribution in AW condition.

Hardness in HAZ in AW condition varies from lowest near the HAZ/TMAZ boundary and increases towards BM. The lowest HAZ hardness is also the global minima of hardness in weld cross-section (Fig. 4.7A). The knockdown in hardness in HAZ is due to the coarsening of precipitates as explained earlier in the discussion of microstructural evolution. The effect of welding speed on hardness in HAZ showed trends similar to that of WN. Welding speed of 140 mmPM reduced the hardness in HAZ to 105 HV as compared to 108 HV when welding speed of 300 mmPM was used [23]. Overall the hardness across the weld cross-section in AW condition shows a "W" shape.

In another study, Liu et al. [13] demonstrated the use of stationary shoulder for FSW of 2219-T6 alloy. They showed that the hardness distributions across the transverse cross-section of the joints welded at different welding speeds were quite different from the welds made with conventional tool with nonstationary shoulder. Fig. 4.7A and B shows the comparative microhardness. The knockdown in HAZ was eliminated. The global minima in hardness was observed in WN for all the cases (Fig. 4.7B). The hardness in WN was higher for faster welding speeds. The width of minimum hardness region decreased with increase in welding speed due to decrease in heat content which softens the material. Overall, it was demonstrated that the use of stationary tool shoulder can manipulate the knockdown in hardness in HAZ in AW condition. However, this approach led to lower overall hardness values in WN and this knockdown may not be desirable.

Hardness evolution in FSW of O temper result is opposite to the T6 temper results. Xu et al. [20] studied the temperature, microstructure, and mechanical property evolution in FSW of 14-mm-thick 2219 alloy in O temper. The hardness distribution across the weld cross-section and depth of the weld is shown in Fig. 4.8. The hardness in WN was significantly higher as compared to BM. The maximum hardness of 95 HV was obtained at the weld top on the advancing side. The hardness along the depth of the weld was observed to decrease. This was

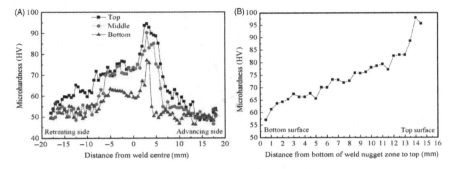

Figure 4.8 Microhardness distribution (A) across the weld transverse cross-section at three different horizontal planes and (B) along the depth in WN [20]. Source: Reprinted with permission from Elsevier.

mainly due to the decrease in peak temperature and cooling rates along the depth of the weld. The core reason for significantly different hardness curve in FSW of O temper is due to the fact that O temper is in annealed state and has lowest strength (50 HV hardness). FSW acts as thermomechanical treatment for the WN and increases strength mainly due to refinement in grain size. Whereas, some degree of precipitation during heating and cooling cycles of FSW result in increased hardness in area surrounding WN (Fig. 4.8A). Note that the hardness maxima in WN shown in Fig. 4.7B is similar to maxima in Fig. 4.8, indicating that WN ends up with similar microstructure.

4.2.2.2 Hardness Evolution in PWHT Condition
Hardness evolution during PWHT is also sensitive to the temper of BM. Liu et al. [23] studied the effect of heat treatment on mechanical properties of friction stir welded joints of 2219-T6 alloy. The hardness distribution across the transverse weld cross-section after PWHT is shown in Fig. 4.9A [23]. Hardness in WN was reported to increase after PWHT. This is mainly due to the strength provided by the precipitation of the θ' precipitates during PWHT as there were solutes available in the matrix due to dissolution during FSW. The highest hardness reported in WN (126 HV) was even higher than the BM hardness (122 HV). However, PWHT could not recover the hardness in HAZ (Fig. 4.9A). In fact, the hardness in HAZ further deteriorated during PWHT. The reduction in hardness in HAZ is attributed to further coarsening during PWHT of already coarsened precipitates in HAZ.

As discussed earlier, FSW of O temper results in contrasting results, similarly hardness evolution during PWHT of FSW of O temper is also quite different. Chen et al. [28] studied the effect of PWHT on the

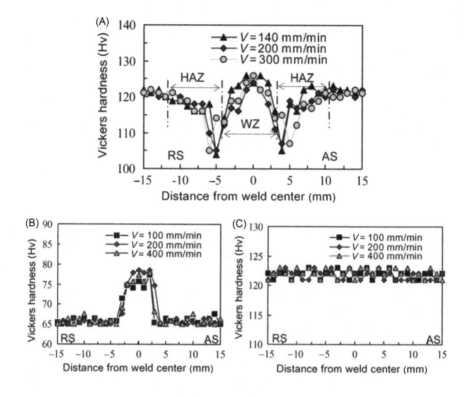

Figure 4.9 Hardness distribution across the weld cross-section in (A) PWHT condition of welds made with differ-ent welding speeds on 2219 alloy in T6 temper [23], (B) AW condition of welds made with different welding speeds on 2219 alloy in O temper [28], and (C) PWHT condition of results shown in (B) [28]. Source: (A) Reprinted with permission from Taylor & Francis. (B) and (C) Reprinted with permission from Springer.

mechanical properties of 2219-O friction stir welded joints. The hardness measurements reported by Chen et al. [28] in AW and PWHT conditions are shown in Fig. 4.9B and C. Hardness evolution in AW condition has already been discussed earlier in Section 4.2.2.1. During PWHT of welds of O temper, hardness increases throughout the weld cross-section and shows a uniform distribution of hardness around 122 HV (Fig. 4.9C), which is equivalent to peak hardness in T6 temper of 2219 alloy. As dis-cussed earlier, the main reason for full recovery of hardness in welds of O temper during PWHT is that the O temper is the weakest temper. Basically, O temper represents the base strength of an alloy and cannot be further degraded during thermomechanical processes such as FSW. Although detailed microstrucutural analysis of precipitates has not been reported, full precipitation of strengthening phase θ' is anticipated as the key reason for full recovery of hardness in the case of FSW of O temper.

4.2.2.3 Tensile Properties

Tensile strength of a FSW joint of Al−Cu alloys also depends on the process parameters, workpiece thickness, and initial temper of the material. FSW results in composite microstructure with varying properties across the weld cross-section. In most of the cases, the region with lowest hardness is the fracture location during tensile test. However, width of the weakest zone can play important role in the mechanics of tensile response. In case of Al−Cu alloys, tensile behavior of FSWed joints has been evaluated in a number of studies [14,18−22,28,31]. Tensile properties of FSWed 2219 alloy from a few studies have been summarized in Table 4.2. The joint efficiency (JE), representing the ratio between the tensile strength of a welded joint and the tensile strength of BM, varies from 69% to 80% except two cases where the defective welds lead to lower JE. It is clear from

Table 4.2 Tensile Properties of FSWed Joints of 2219-T6 Alloy Reported in the Literature							
RPM	IPM	W_T	BM		Weld-AW		References
			YS	UTS	UTS	JE	
800	4	5.8	315	416	295.3	71	[21]
900	4	5.8	315	416	301.6	72	[21]
1000	4	5.8	315	416	316.1	76	[21]
1100	4	5.8	315	416	328.6	79	[21]
1200	4	5.8	315	416	320.3	77	[21]
1300	4	5.8	315	416	212.1	51	[21]
800	5.5	5.8	315	416	289.1	69	[21]
900	5.5	5.8	315	416	293.2	70	[21]
1000	5.5	5.8	315	416	312	75	[21]
1100	5.5	5.8	315	416	323.6	78	[21]
1200	5.5	5.8	315	416	318.2	76	[21]
1300	5.5	5.8	315	416	174.3	42	[21]
800	4	7.4	315	432	324	75	[32]
800	4	7.4	315	432	346	80	[32]
800	2	7.4		432	318	73	[16]
800	4	7.4		432	339	78	[16]
800	6	7.4		432	347	80	[16]
800	8	7.4		432	178	41	[16]

RPM, rotation per minute; IPM, inches per minute; W_T, weld thickness in mm; YS, yield strength in MPa; UTS, ultimate tensile strength in MPa; JE, joint efficiency in %.

Table 4.2, increase in tool rotation rate increases the tensile strength of the joint. Higher tool rotation rate translates to higher heat input during welding which results in better degree of solutionization. Lower heat input welds can result in formation of coarse grain boundary phases which can severely impact the strength and ductility of the joint.

Liu et al. [23] studied the effect of heat treatment on tensile properties of FSWed joints of 2219-T6. They also investigated the effect of welding speed on tensile properties, using a constant tool rotation rate (800 RPM). The results of tensile test of joints in AW and PWHT conditions are shown in Fig. 4.10. It is clear from the results shown in Fig. 4.10A that the tensile strength of joints in AW condition increases with increasing welding speed, except for the case of 300 mmPM welding speed in which strength dropped drastically due to defects in the weld. The maximum tensile strength of 336 MPa was obtained for the welding speed of 220 mmPM. It is equivalent to 81% of the tensile strength of the base material. The increase in tensile strength in AW condition was correlated to the hardness results shown in Fig. 4.7A. Reduction in knockdown in HAZ strength with increase in welding speed resulted in better tensile properties. Elongation to failure also showed trends similar to tensile strength in AW condition. After PWHT, the tensile strength of the joints increased and reduction in elongation to failure was observed (Fig. 4.10B). The highest joint strength of 89% of BM strength was obtained for defect-free weld made with highest speed (220 mmPM). The recovery of the hardness in WN due to precipitation of θ' precipitates during PWHT resulted in the increase in the tensile strength and the decrease in the elongation at fracture of the heat-treated joints.

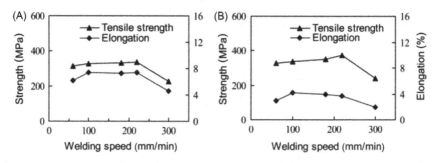

Figure 4.10 Tensile properties of welds of 2219-T6 alloy in (A) AW condition and (B) PWHT condition [23].
Source: Reprinted with permission from Taylor & Francis.

4.3 FSW OF Al—Cu—Mg ALLOYS

Al—Cu—Mg alloys are one of most complex precipitation-strengthened alloy systems. The precipitation sequence in Al—Cu—Mg alloys during artificial aging is

SSSS → solute clusters → solute clusters + GPB zones → solute clusters
+ GPB zones + S → S

It is also the most widely used alloy category in aerospace industry and alloy 2024 is a widely used principal damage-tolerant material in the aerospace industry in wing and fuselage structures. The structures made from alloy 2024 are usually fastened together using rivets due to its poor weldability by fusion processes as described in detail in earlier chapters. A large number of studies on various aspects of joining of alloy 2024 using FSW technique exist [27,33−43]. Majority of the studies have been conducted on T3 temper, but T4 and T6 tempers have also been studied. For precipitation-hardened aluminum alloys (2XXX, 6XXX, and 7XXX), the thermomechanical processing during FSW results in the change in the distribution, size, and density of the precipitates creating various zones as described earlier in the previous chapters. The peak temperature in the WN is sufficiently high to induce the dissolution of the strengthening precipitates. TMAZ forms on the both sides at the boundary of WN towards BM. It experiences lower heat input and smaller plastic deformation. HAZ, which is subjected only to the thermal cycle, generally forms a region with low hardness. Depending upon the thermal stability of precipitates, HAZ mainly forms due to coarsening of precipitates. Dissolution of low thermal stability phases such as δ' (Al$_3$Li) can also occur in HAZ. Zhang et al. [33], in a detailed study on hardness evolution in HAZ during natural aging of FSW of 2024-T351 alloy, termed the low hardness region into low hardness zone I (LHZ I) and LHZ II. This was done to explain the complex microstructural variation arising in the weld cross-section of 2024-T351 alloy during FSW. It will be discussed further in greater detail in upcoming sections.

4.3.1 Evolution of Microstructure and Hardness

Microstructural evolution in various zones in FSW of alloy 2024-T3 alloy is relatively complex and can be better understood by correlating to hardness evolution across the transverse cross-section of the weld. Fig. 4.11 shows typical hardness profiles along the transverse

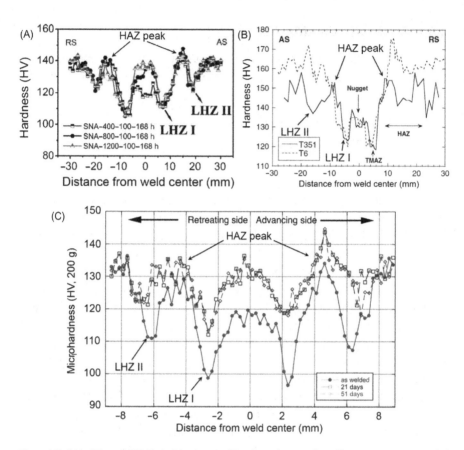

Figure 4.11 (A), (B), and (C) Typical hardness profiles observed across the weld transverse cross-section of FSW of 2024 alloy in T3 temper [33,40,45]. Source: (A) and (B) Reprinted with permission from Elsevier. (C) Reprinted with permission from Dr. A.K. Shukla.

cross-section of FSW of 2024 alloy in T3 temper [33,40,45]. WN shows typical lower hardness compared to BM. HAZ and TMAZ produce complex microstructural variation. The lowest hardness in AW condition is identified in TMAZ in a few studies [41,44,45], whereas majority of the studies [33,35,37,39,40,46] acknowledged a region in HAZ which is closer to TMAZ as the lowest hardness region. Therefore, in this section the discussion of weakest region in FSW of 2024-T3 alloy will be correlated to the HAZ. From the hardness profiles shown in Fig. 4.11, three distinct areas in the HAZ can be identified:

1. A hardness minimum in the HAZ just outside the TMAZ—adapted from Zhang et al. [33], this region will be called low hardness zone I (LHZ I) as labeled in Fig. 4.11.

2. Away from WN, LHZ I is followed by a hardness peak in the HAZ—this region will be called HAZ peak as labeled in Fig. 4.11.
3. A second hardness minimum farther away in the HAZ—this region will be called low hardness zone II (LHZ II) as labeled in Fig. 4.11.

LHZ I is the global minima in hardness profile and is usually the weakest region. LHZ I and LHZ II form on both the sides of the WN.

As discussed earlier in the physical metallurgy section, Al—Cu—Mg alloys are strengthened by Cu—Mg co-clusters, GPB zones, and S precipitates. Alloys in T3 or T4 temper are strengthened by Cu—Mg co-clusters and GPB zones whereas S precipitates dominate in aged tempers (T6 or T8). Many researchers have explained the microstructural evolution in various zones of FSW of 2024 alloy in various tempers [33—46].

4.3.1.1 Weld Nugget

WN of FSW of 2024 alloy experiences high temperature in the range of 450—550°C. Fig. 4.12 shows the TTT diagram for precipitates in Al—Cu—Mg alloy system. It is clear from TTT diagram (Fig. 4.12)

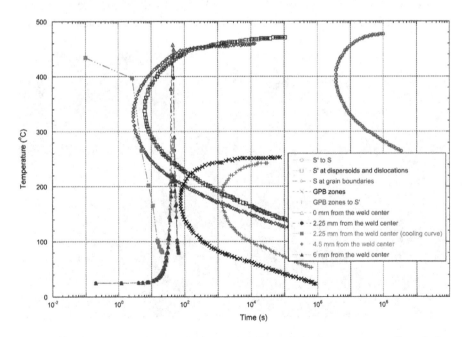

Figure 4.12 TTT diagram of various second phases in 2024 alloy. Note that the temperature profiles and other legends are explained in [40]. Source: Reprinted with permission from Dr. A.K. Shukla.

and Table 4.1 that all the second phases in 2024 alloy dissolve upon exposure to high temperatures reached in WN during FSW. Fig. 4.12 also shows the numerically calculated temperature profile of WN by Shukla [40] superimposed on the TTT diagram. It clearly shows that the temperature in WN is high enough for dissolution of strengthening phases. Therefore, GPB zones and solute clusters in underaged temper and S precipitates in aged tempers dissolve under such high temperatures and large plastic deformation in the WN during FSW. Shukla [40] studied FSW of thin sheets of alloy 2024 in T3 temper. Shukla demonstrated that the WN showed equiaxed, recrystallized grains with size less than 4 μm, and grain structure with very low dislocation density as shown in Fig. 4.13.

Fig. 4.13A shows a distribution of $Al_{20}Cu_2Mn_3$ dispersoids and reprecipitated phases during cooling cycle of FSW. Reprecipitation during cooling cycle can occur as discussed earlier in Section 4.2.1.1. The microstructure in WN also exhibits a low dislocation density of dislocations as shown in high magnification TEM image (Fig. 4.13B). A few helical dislocation structures were also observed in WN in the AW condition as shown in Fig. 4.13. These helical dislocation structures form during cooling cycle of FSW due to dynamic recrystallization process. Jones et al. [46] also reported similar characteristics in

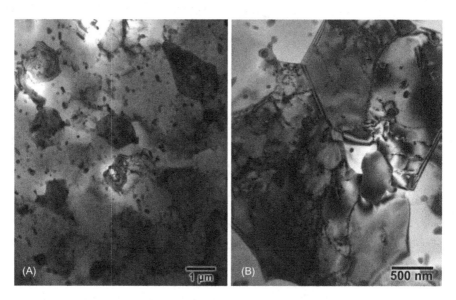

Figure 4.13 (A) Low magnification and (B) high magnification TEM images taken along [0 0 1] zone axis of WN of FSWed 2024-T3 alloy [40]. Source: Reprinted with permission from Dr. A.K. Shukla.

WN of FSWed 2024-T351 alloy in the AW condition. They confirmed presence of fine S precipitates in the SADP taken along [0 0 1] zone axis of a sample from WN [46]. Dixit et al. [39] also confirmed these findings in their study on the effect of initial temper on mechanical properties of FSWed 2024 alloy. Dixit et al. [39] observed and concluded that initial temper of the material is insignificant for the microstructural evolution in WN, provided the welding parameters are similar. This is due to the fact that temperature in the WN is high enough for dissolution of GPB zones in underaged temper and S precipitates in peak aged temper which result in similar microstructure. However, reprecipitation during cooling cycle can play significant role in determining the final microstructure in the WN in the AW condition. Cooling after FSW is dependent on welding speed and auxiliary cooling methods used during the process as discussed in earlier chapters.

PWHT can further alter the microstructure in WN of FSWed 2024 alloy. Chen et al. [27] reported the evolution of microstructure in WN of FSWed 2024-T3 alloy during PWHT. PWHT was carried out at 220°C for 10 hours. Fig. 4.14 shows detailed microstructure in WN in

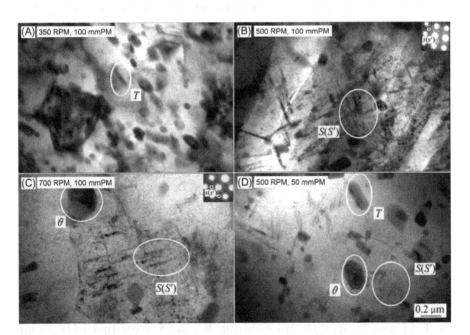

Figure 4.14 (A), (B), (C), and (D) Detailed TEM analysis of microstrucuture of WN of alloy 2024-T3 in PWHT condition. Micrographs are labeled with welding parameters and identified second phases [27]. Source: Reprinted with permission from Elsevier.

the PWHT condition of alloy 2024-T3 welded using different process parameters. The rod-shaped particles were identified as T-phase ($Al_{20}Cu_2Mn_3$), which are dispersoids in the Al–Cu–Mg system and can remain undissolved during FSW because of its high dissolution temperature. The oval-shaped particles were identified as θ phase (Al_2Cu) which are rich in Al and Cu atoms and formed on grain boundaries and interior during cooling of the weld. SADP shown in the inset of Fig. 4.14B confirms the presence of S precipitate formed during PWHT. A few coarsened S phases were also observed as shown in Fig. 4.14A, indicating some degree of coarsening of phases took place during PWHT. However, this may also be related to the coarsening of reprecipitated phases formed during cooling of the weld cycle. Absence of GPB zones in SADP also indicates that most of the GP zones transform to S phase during PWHT. Chen et al. [27] observed that maximum S phase precipitation occurred in a weld with high heat input and high welding speed (500 RPM, 100 mmPM) as shown in Fig. 4.14B. Overall, similar to Al–Cu alloy system, WN of FSWed Al–Cu–Mg system results in partial to full dissolution of strengthening phases during FSW. Cooling cycle of the weld also plays a significant role in determining the quantity of retained solutes in WN for precipitation of strengthening phases during PWHT.

4.3.1.2 Heat Affected Zone—Low Hardness Zones

As described earlier, HAZ is the most critical zone that evolves during FSW of 2024 alloy. Two low hardness zones form due to various metallurgical phenomena occurring in the complex system of Al–Cu–Mg alloys. In precipitation-strengthened aluminum alloys, the hardness profile greatly depends on the precipitate distribution and only slightly on the grain size and dislocation structures. Quantitative analysis of the volume fraction, type, and size of the precipitates can explain the microstructural evolution. Genevois et al. [45] performed a quantitative investigation of microstructural evolution in HAZ of FSWed of 2024-T351 and T6 using TEM, SAXS (small angle X-ray scattering), and DSC. They calculated the volume fraction of precipitates by the combination of DSC and SAXS techniques, and the precipitate size by combining TEM and SAXS analyses. Volume fraction of GPB zones and S precipitates in BM and various zones of the welds of 2024 alloy in T351 and T6 tempers are shown in Fig. 4.15. Note that LHZ I and LHZ II are not labeled for the weld of T6 temper as only one LHZ forms in the case of peak aged material known as HAZ. BM of T6

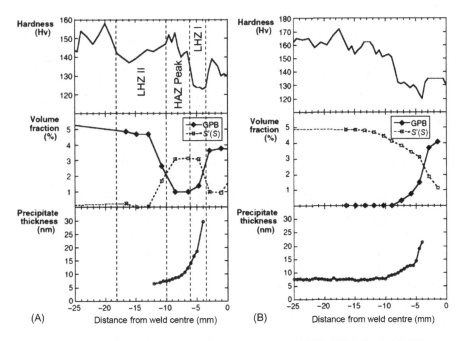

Figure 4.15 Quantitative details of precipitate evolution in various zones of FSWed 2024 alloy in (A) T351 temper and (B) T6 temper [45]. Source: Reprinted with permission from Elsevier.

temper showed high volume fraction (5%) of S precipitates whereas T351 temper showed presence of only GPB zones with a volume fraction of 5.5%. This evaluation also revealed that the WN showed similar volume fractions of GPB zones and S phases, which confirms that the initial temper of the base metal does not influence the hardness and microstructure of WN after welding. Microstructural evolution in the HAZ is more complex because different precipitate distribution across the weld cross-section evolves due to temperature gradient of FSW process. In T351 weld, GPB zones are dominant in BM but their volume fraction reduces as one approaches LHZ II. LHZ II is characterized by the dissolution of the GPB zones. During welding, the gradual increase in temperature causes the beginning of the GPB zone reversion and results in hardness reduction in LHZ II zone. Zhang et al. [33] experimentally showed that temperature in LHZ II can reach up to 200–220°C. As shown in Table 4.1, such high temperatures can result in partial dissolution of the GPB zones and solute clusters in the LHZ II. Zhang et al. [33] also concluded that LHZ II formed due to dissolution of GPB zones and solute clusters.

Moving further closer from LHZ II zone to WN, HAZ peak zone forms with peak hardness similar or higher than BM T351 hardness. In this region, the volume fraction of GPB zones falls sharply and S phase fraction reaches its maximum as shown in Fig. 4.15. Temperature in this zone is expected to be higher than 200°C (LHZ II zone temperature) but lower than 300°C, which falls in LHZ I zone temperature. As per Table 4.1 and Fig. 4.12, this intermediate temperature (200−300°C) exposure for a short time is appropriate enough for precipitation of GPB zones and fine-scale S phase and mimics an artificial aging heat treatment, which can explain the strengthening in this zone. Genevois et al. [45] in their quantitative analysis observed the formation of higher volume fraction of fine S precipitates and a relatively low fraction of GPB zones for the peak in hardness. Jones et al. [46] also observed fine S precipitates in HAZ peak region as shown in Fig. 4.16C. Whereas, Shukla [40] attributed the peak in HAZ hardness to formation of only GPB zones. Therefore, precipitation of S phase and GPB zones in HAZ peak region due to appropriate combination of temperature and time can be ascertained as key reason for increased hardness.

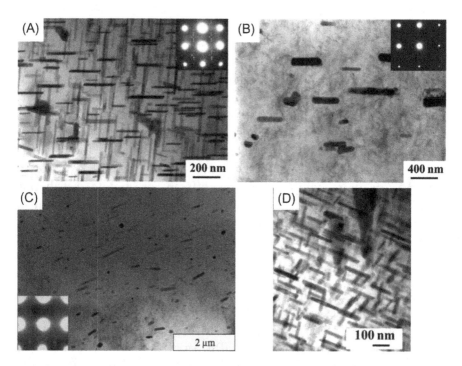

Figure 4.16 Microstructural details of (A) LHZ I [33], (B) LHZ II [33] and (C) HAZ peak zones in AW condition [46], and (D) LHZ I in PWHT condition [45]. Source: Reprinted with permission from Elsevier.

LHZ I zone forms at the boundary region between HAZ and TMAZ. LHZ I experiences temperature above 330°C as experimentally demonstrated by Zhang et al. [33]. Such high temperatures fall under the overaging regime of S precipitates and even exposure for short duration, as in the case of FSW, can result in high level of coarsening of precipitates. Fig. 4.15A shows that the volume fraction of S phase and GPB zone in LHZ I is similar to that of HAZ peak zone and increases as one moves towards WN. However, as reported by Genevois et al. [45], the thickness of S precipitate was observed to increase from 5 to 8 nm in HAZ peak towards 30 nm in LHZ I, which confirms that coarse S precipitates form in this zone. TEM image of a sample from LHZ I is shown in Fig. 4.16 [33] also confirms the presence of high density of coarse S precipitates in LHZ I during FSW and consequent reduction in hardness. Genevois et al. [45] also studied the effect of PWHT on hardness and microstructural evolution in LHZ I of FSWed 2024-T351 alloy. Standard T6 aging treatment for 10 hours at 190°C was chosen as the PWHT, which transforms the GPB zones to fine S precipitates in BM of T351 material. Genevois et al. [45] observed that away from the WN, the PWHT increases the hardness of the 2024-T351 weld up to that of the 2024-T6. LHZ II was also noted to have increased hardness to the level of T6 BM hardness, due to precipitation of fine S precipitates. However, as shown in a TEM image in Fig. 4.16A, LHZ I exhibited extremely coarsened S precipitates. These coarse precipitates resulted in further reduction in hardness of LHZ I during PWHT.

FSW of 2024 alloy in aged tempers (T6 or T8) results in contrasting microstructure. Genevois et al. [45] also reported a quantitative analysis of microstructural evolution in various zones in a FSWed 2024-T6 alloy. The results of the study are shown in Fig. 4.15B. It is clear from Fig. 4.15B that WN of weld of T6 temper exhibited similar volume fraction of GPB zones and S precipitates, and thus showed similar hardness in AW condition. Whereas, weld of T6 temper did not show the peculiar splitting of HAZ into low hardness zones and HAZ peak as observed in welds of T3 temper. This is due to the fact that T6 temper is strenghened by S precipitates which have higher thermal stability as compared to GPB zones and co-clusters. Thus, reversion and peak aging phenonmon that occurs in LHZ II and HAZ peak zones of T3 temper weld cannot occur in the case of T6 temper weld as the material is already aged to a stable condition. However, the region where

LHZ I was observed in T3 temper weld experiences high temepratures in the range of 300−350°C. Even short duration of such high temperatures is enough for coarsening of already existing S precipitates. Genevois et al. [45] reported the precipitate thickness across the weld cross-section (Fig. 4.15B) by combined analysis of SAXS and DSC experiments. The precipitates were three times thicker in HAZ (LHZ I) as compared to BM of T6 temper. It shows that thermal cycle in HAZ during FSW resulted in extensive coarsening of S precipitates and resultant reduction in the hardness.

4.3.1.3 Welding Parameters and Natural Aging Response

FSW parameters are important factor in determining the final microstructure and hardness in FSW of 2024 alloy in various tempers. Generally, increase in tool rotation rate and decrease in welding speed result in higher heat content during the welding. Understandably, decrease in tool rotation rate and increase in welding speed result in lower heat content during the welding. Tool rotation rate has greater influence on peak temperature during welding, whereas welding speed affects the heating and cooling rates during welding.

The 2024 alloy has a strong natural aging tendency. As discussed in earlier chapters, the 2024 alloy is strengthened by Cu−Mg co-clusters and GPB zones. Al−Cu−Mg alloys achieve a major fraction of its peak strength in naturally aged (underaged) condition. Due to this reason, 2024 alloy has many applications in underaged condition. Zhang et al. [33] studied the effect of process parameters and natural aging response of weld of 2024-T351 alloy. Zhang et al. [33] used a combination of tool rotation rate of 400, 800, 1200 RPM and welding speed of 100, 200, 400 mmPM as welding parameters. Zhang et al. [33] carried out hardness measurement across the weld cross-section after 24, 96, 168, 2880, 8760 hours of natural aging. The hardness measurement plots in Fig. 4.17 show the effect of welding parameters and natural aging of hardness evolution across the weld cross-section. Fig. 4.17A shows the effect of tool rotation rate on hardness across the weld cross-section. A notable difference in hardness can be observed in WN of three plots shown in Fig. 4.17A where sample with higher tool rotation rate showed higher hardness. As discussed earlier, higher tool rotation rate results in higher temperature and thus leads to higher degree of solutionization in WN during FSW which provides higher solute content for co-cluster and GPB formation during natural aging

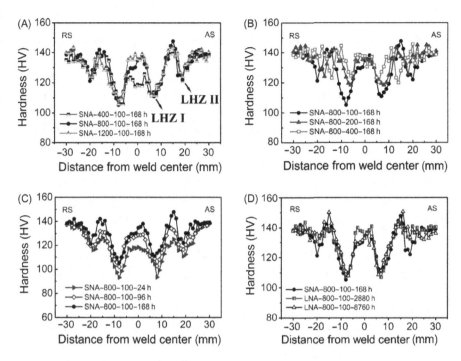

Figure 4.17 Hardness results showing the effect of (A) and (B) welding parameters and (C) and (D) natural aging across the weld cross-section of 2024-T351 alloy [33]. Source: Reprinted with permission from Elsevier.

in WN. Interestingly, no significant effect of tool rotation rate on hardness drop in LHZ I and LHZ II was observed. Zhang et al. [33] concluded that for all the welds shown in Fig. 4.17A, the temperature in LHZ I was high enough to promote extensive coarsening of S phase, which resulted in hardness drop in all the welds. LHZ II zone which experiences reversion treatment and shows full dissolution of GPB zones also showed insensitivity to the tool rotation rate. Fig. 4.17B shows the effect of welding speed with a constant tool rotation rate on hardness in welds of 2024 alloy in T351 temper. Since tool rotation rate was constant, it is understandable that hardness in WN was similar in all three welds as shown in Fig. 4.17B. Welding speed has larger impact on cooling rates, thus the extent of coarsening of S phase in LHZ I will be dependent on welding speed. Also, the distance of LHZ I with respect to WN also changes with change in welding speed. It is evident from Fig. 4.17B that lower welding speed resulted in higher drop in hardness in LHZs which is due to higher residence time at high temperature resulting in extensive coarsening in LHZ I and dissolution in LHZ II. Similarly, the location of LHZ II showed similar trends with change in welding speed. However, the drop in hardness in LHZ II was observed to be insensitive to the change in welding speed. This is due to the fact that LHZ II forms due to the dissolution of clusters and GPB zones which has extremely fast kinetics at high temperatures. Thus, it is safe to conclude that the formation of LHZ II zone is avoidable under traditional FSW parameters. Employment of external cooling medium can be utilized to manipulate the formation of LHZ II. Fig. 4.17C and D shows the effect of natural aging on hardness of weld made with a single set of welding parameters. It can be observed that the hardness showed significant improvement in both LHZ I and LHZ II during initial natural aging for 168 hours (Fig. 4.17C). However, later stage of long-term (2880 hours) natural aging did not show any significant improvement in hardness in LHZ I, but increased and approached the BM level in LHZ II. The HAZ peak also showed considerable increase in hardness during the entire natural aging regime as can be seen in Fig. 4.17C and D. Zhang et al. [33] experimentally observed and attributed the slow recovery of hardness in LHZ II to the reformation of Cu–Mg cluster during natural aging. Whereas the low level of hardness increase in LHZ I was attributed to the nonavailability of solutes to form any clusters or GPB zones during natural aging as coarsening in LHZ I during FSW consumes a high fraction of solutes from matrix.

4.3.2 Mechanical Behavior

FSW of Al–Cu–Mg alloys results in four to five different zones (BM, WN, TMAZ, HAZ, or LHZ I and LHZ II) across the weld cross-section depending on the initial temper of the material. All these zones have strikingly different microstructural characteristics as explained in detail in pervious sections. This results in a complex composite weld cross-section containing a total of seven to nine different microstructural zones in the gage section. The global mechanical behavior of the weld depends on the local properties across these zones, which are dependent on the local microstructure produced during welding. Therefore, it is important to understand the local mechanical response of different zones in order to determine the global weld strength and ductility.

Genevois et al. [41] studied the effect of the microstructural heterogeneity on the global and local tensile properties of welds of alloy 2024 in T351 and T6 temper. They investigated local mechanical properties by conducting tensile tests on microspecimens selected from various locations across the weld. They also used digital image correlation (DIC) technique during transverse tensile tests of macrosized samples to obtain local and global stress–strain curves. DIC is a powerful technique in determining the displacement of an object and hence can be used to capture deformation fields developed during a tensile test. The specimen is prepared by applying a random speckle pattern on the surface to be studied, generally using black and white spray paints. During a tensile test, a camera is placed normal to the specimen plane and a reference image of the undeformed specimen is acquired. Later, a series of time-based images are captured during the test. A computer algorithm compares the local displacement of speckled pattern with the reference image and generates a displacement plot.

The local mechanical properties of welds of 2024 alloy measured by Genevois et al. [41] are shown in Fig. 4.18. The difference in microstructure of weld of T351 and weld of T6 temper can help in understanding the distribution of local mechanical properties. The local yield strength (YS) profile of the 2024-T351weld (Fig. 4.18A) shows a trend similar to the hardness profile discussed in earlier sections. The lowest YS is found in LHZ I close to TMAZ. The WN shows a slightly higher strength mainly due to the natural aging after welding. High strength in HAZ peak region and another low strength region which forms LHZ II is also evident from the YS plot. The localized ultimate tensile strength (UTS) showed a response similar to that of YS, except that the differences between the UTS of different zones are

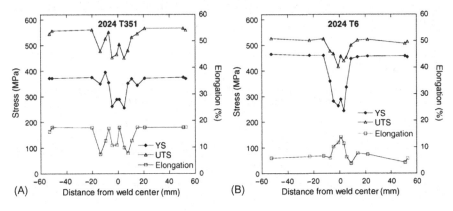

Figure 4.18 Local mechanical properties of various zones in welds of 2024 alloy in (A) T351 and (B) T6 tempers [41].
Source: Reprinted with permission from Elsevier.

smaller in magnitude. The local ductility was observed to be lowest in the HAZ peak area. This is due to the fact that HAZ peak is microstructurally closer to peak aged material which usually show lower ductility as compared to T351 temper due to increase in number of obstacles to dislocation motion as a result of precipitation.

In case of weld of 2024-T6 alloy (Fig. 4.18B), the lowest strength was observed in WN and HAZ (closer to TMAZ). It directly reflects the hardness characteristics of the weld of peak aged alloy. The local elongation was observed to be highest in WN which is due to the formation of solid solution in WN during FSW. Extensive coarsening of S precipitates in HAZ leads to reduction in both local strength and ductility. Overall, Genevois et al. [41] observed that both the welds of alloy 2024 showed a very high hardening rate and a low ductility. The low ductility is due to the strain localization in the weaker zones of the weld. In case of weld of T6 material, severe strain localization can reduce elongation to very low values, much lower than would be expected for aluminum alloys as the difference between the local strength of weakest zone and BM is much higher as compared to the case of T3 material weld. Strain localization in transverse tensile tests of macrosample can influence the measured strength. This is due to the fact that necking and failure will occur within the weakest location. Genevois et al. [41] demonstrated that for the T3 material weld, the deformation localizes in the LHZ I (close to TMAZ) and the WN. The final fracture initiates between these two zones. Also, a smaller degree of strain localization was also observed in LHZ II by Genevois et al. [41], where clusters and GPB zones dissolves during FSW. Overall, early strain localization results in much lower strength and ductility in both the welds.

REFERENCES

[1] M. Starink, Analysis of aluminium based alloys by calorimetry: quantitative analysis of reactions and reaction kinetics, Int. Mater. Rev. 49 (2004) 191—226.

[2] H. Sidhar, N.Y. Martinez, R.S. Mishra, J. Silvanus, Friction stir welding of Al—Mg—Li 1424 alloy, Mater. Des 106 (2016) 146—152.

[3] H. Sidhar, R.S. Mishra, Aging kinetics of friction stir welded Al—Cu—Li—Mg—Ag and Al—Cu—Li—Mg alloys, Mater. Des 110 (2016) 60—71.

[4] J.M. Papazian, A calorimetric study of precipitation in aluminum alloy 2219, Metal. Trans. A 12 (1981) 269—280.

[5] G.E. Totten, D.S. MacKenzie, Handbook of Aluminum: Vol. 1: Physical Metallurgy and Processes, CRC Press, USA, 2003.

[6] S. Wang, M. Starink, Precipitates and intermetallic phases in precipitation hardening Al—Cu—Mg—(Li) based alloys, Int. Mater. Rev. 50 (2005) 193—215.

[7] N.E. Prasad, A. Gokhale, R. Wanhill, Aluminum—Lithium Alloys: Processing, Properties, and Applications, Butterworth-Heinemann, UK, 2013.

[8] B. Malard, F. De Geuser, A. Deschamps, Microstructure distribution in an AA2050 T34 friction stir weld and its evolution during post-welding heat treatment, Acta Mater 101 (2015) 90—100.

[9] J. Li, H. Liu, Optimization of welding parameters for the reverse dual-rotation friction stir welding of a high-strength aluminum alloy 2219-T6, Int. J. Adv. Manuf. Technol. 76 (2015) 1469—1478.

[10] W. Xu, J. Liu, D. Chen, G. Luan, Low-cycle fatigue of a friction stir welded 2219-T62 aluminum alloy at different welding parameters and cooling conditions, Int. J. Adv. Manuf. Technol. 74 (2014) 209—218.

[11] X. Feng, H. Liu, J.C. Lippold, Microstructure characterization of the stir zone of submerged friction stir processed aluminum alloy 2219, Mater. Charact 82 (2013) 97—102.

[12] H. Liu, X. Feng, Effect of post-processing heat treatment on microstructure and microhardness of water-submerged friction stir processed 2219-T6 aluminum alloy, Mater. Des 47 (2013) 101—105.

[13] H. Liu, J. Li, W. Duan, Friction stir welding characteristics of 2219-T6 aluminum alloy assisted by external non-rotational shoulder, Int. J. Adv. Manuf. Technol. 64 (2013) 1685—1694.

[14] H. Liu, H. Zhang, Q. Pan, L. Yu, Effect of friction stir welding parameters on microstructural characteristics and mechanical properties of 2219-T6 aluminum alloy joints, Int. J. Mater. Form. 5 (2012) 235—241.

[15] H. Liu, H. Zhang, L. Yu, Homogeneity of mechanical properties of underwater friction stir welded 2219-T6 aluminum alloy, J. Mater. Eng. Perform 20 (2011) 1419—1422.

[16] H. Liu, H. Zhang, L. Yu, Effect of welding speed on microstructures and mechanical properties of underwater friction stir welded 2219 aluminum alloy, Mater. Des 32 (2011) 1548—1553.

[17] H. Zhang, H. Liu, L. Yu, Microstructural evolution and its effect on mechanical performance of joint in underwater friction stir welded 2219-T6 aluminium alloy, Sci. Technol. Weld. Join. 16 (2011) 459—464.

[18] K.S. Arora, S. Pandey, M. Schaper, R. Kumar, Effect of process parameters on friction stir welding of aluminum alloy 2219-T87, Int. J. Adv. Manuf. Technol. 50 (2010) 941—952.

[19] Y. Chen, J. Feng, H. Liu, Precipitate evolution in friction stir welding of 2219-T6 aluminum alloys, Mater. Charact 60 (2009) 476—481.

[20] W. Xu, J. Liu, G. Luan, C. Dong, Temperature evolution, microstructure and mechanical properties of friction stir welded thick 2219-O aluminum alloy joints, Mater. Des 30 (2009) 1886—1893.

[21] W. Xu, J. Liu, G. Luan, C. Dong, Microstructure and mechanical properties of friction stir welded joints in 2219-T6 aluminum alloy, Mater. Des 30 (2009) 3460—3467.

[22] Y. Chen, H. Liu, J. Feng, Friction stir welding characteristics of different heat-treated-state 2219 aluminum alloy plates, Mater. Sci. Eng. A 420 (2006) 21—25.

[23] H. Liu, Y. Chen, J. Feng, Effect of heat treatment on tensile properties of friction stir welded joints of 2219-T6 aluminium alloy, Mater. Sci. Technol. 22 (2006) 237—241.

[24] R.S. Mishra, P.S. De, N. Kumar, Friction Stir Welding and Processing: Science and Engineering, Springer, USA, 2014.

[25] R.S. Mishra, Z. Ma, Friction stir welding and processing, Mater. Sci. Eng. R Rep. 50 (2005) 1–78.

[26] M.F. Ashby, D. Jones, Engineering Materials 2: An Introduction to Microstructures, Processing and Design, Elsevier, Netherlands, 2014.

[27] C. Yu, D. Hua, J. Li, J. Zhao, M. Fu, X. Li, Effect of welding heat input and post-welded heat treatment on hardness of stir zone for friction stir-welded 2024-T3 aluminum alloy, Trans. Nonferrous Metals Soc. China 25 (2015) 2524–2532.

[28] Y. Chen, H. Liu, J. Feng, Effect of post-weld heat treatment on the mechanical properties of 2219-O friction stir welded joints, J. Mater. Sci. 41 (2006) 297–299.

[29] W. Xu, J. Liu, Microstructure and pitting corrosion of friction stir welded joints in 2219-O aluminum alloy thick plate, Corros. Sci. 51 (2009) 2743–2751.

[30] Y. Chen, J. Feng, H. Liu, Stability of the grain structure in 2219-O aluminum alloy friction stir welds during solution treatment, Mater. Charact 58 (2007) 174–178.

[31] C. Lee, D. Choi, W. Lee, S. Park, Y. Yeon, S. Jung, Microstructures and mechanical properties of double-friction stir welded 2219 Al alloy, Mater. Trans. 49 (2008) 885–888.

[32] H. Liu, H. Zhang, Y. Huang, Y. Lei, Mechanical properties of underwater friction stir welded 2219 aluminum alloy, Trans. Nonferrous Metals Soc. China 20 (2010) 1387–1391.

[33] Z. Zhang, B. Xiao, Z. Ma, Hardness recovery mechanism in the heat-affected zone during long-term natural aging and its influence on the mechanical properties and fracture behavior of friction stir welded 2024Al–T351 joints, Acta Mater. 73 (2014) 227–239.

[34] Z. Zhang, W. Li, Y. Feng, J. Li, Y. Chao, Improving mechanical properties of friction stir welded AA2024-T3 joints by using a composite backplate, Mater. Sci. Eng. A 598 (2014) 312–318.

[35] Z. Zhang, W. Li, J. Li, Y. Chao, Effective predictions of ultimate tensile strength, peak temperature and grain size of friction stir welded AA2024 alloy joints, Int. J. Adv. Manuf. Technol. 73 (2014) 1213–1218.

[36] A. Eramah, M.P. Rakin, D.M. Velji, A. Nenad, M. Zrilic, M.P. Milenko, Influence of friction stir welding parameters on properties of 2024 T3 aluminium alloy joints, Thermal Sci. 17 (2013) 21–27.

[37] R. Fu, J. Zhang, Y. Li, J. Kang, H. Liu, F. Zhang, Effect of welding heat input and post-welding natural aging on hardness of stir zone for friction stir-welded 2024-T3 aluminum alloy thin-sheet, Mater. Sci. Eng. A 559 (2013) 319–324.

[38] V. Dixit, R. Mishra, R. Lederich, R. Talwar, Influence of process parameters on microstructural evolution and mechanical properties in friction stirred Al-2024 (T3) alloy, Sci. Technol. Weld. Join. 14 (2009) 346–355.

[39] V. Dixit, R. Mishra, R. Lederich, R. Talwar, Effect of initial temper on mechanical properties of friction stir welded Al-2024 alloy, Sci. Technol. Weld. Join. 12 (2007) 334–340.

[40] A.K. Shukla, Friction stir welding of thin-sheet, age-hardenable aluminum alloys: a study of process/structure/property relationships, ProQuest Dissertations and Theses (2007).

[41] C. Genevois, A. Deschamps, P. Vacher, Comparative study on local and global mechanical properties of 2024 T351, 2024 T6 and 5251 O friction stir welds, Mater. Sci. Eng. A 415 (2006) 162–170.

[42] D.G. Moghadam, K. Farhangdoost, R.M. Nejad, Microstructure and residual stress distributions under the influence of welding speed in friction stir welded 2024 aluminum alloy, Metal. Mater. Trans. B 47 (2016) 2048–2062.

[43] N. Nadammal, S.V. Kailas, S. Suwas, A bottom-up approach for optimization of friction stir processing parameters; a study on aluminium 2024-T3 alloy, Mater. Des 65 (2015) 127–138.

[44] C. Genevois, D. Fabrègue, A. Deschamps, W.J. Poole, On the coupling between precipitation and plastic deformation in relation with friction stir welding of AA2024 T3 aluminium alloy, Mater. Sci. Eng. A 441 (2006) 39–48.

[45] C. Genevois, A. Deschamps, A. Denquin, B. Doisneau-Cottignies, Quantitative investigation of precipitation and mechanical behaviour for AA2024 friction stir welds, Acta Mater 53 (2005) 2447–2458.

[46] M. Jones, P. Heurtier, C. Desrayaud, F. Montheillet, D. Alléhaux, J. Driver, Correlation between microstructure and microhardness in a friction stir welded 2024 aluminium alloy, Scr. Mater. 52 (2005) 693–697.

CHAPTER 5

Friction Stir Welding of Al–Li Alloys

5.1 INTRODUCTION

Al–Li alloys have regained significant attention from the aerospace industry due to their higher specific strength compared to conventional aerospace aluminum (2XXX and 7XXX) alloys. The third generation of Al–Li alloys is considered as promising candidates for various components of aircrafts due to their excellent combination of low density, high strength, high stiffness, and excellent damage tolerance [1]. Industrially, Al–Li alloys can be divided into two subcategories: Al–Mg–Li alloys and Al–Cu–Li alloys. Al–Mg–Li alloys are less popular mainly due to low thermal stability and moderate strength. Al–Cu–Li alloys are mainly utilized in T3 or T8 tempers due to the enhanced precipitation of T_1 precipitate in presence of dislocations. Other aspects of physical metallurgy and microstructural characteristics of Al–Li alloys have been discussed in earlier chapters.

Friction stir welding (FSW) of various precipitation-strengthened aluminum alloys results in a variety of microstructure. Of course, initial microstructure and alloy chemistry plays the most important role in determining the final microstructure. Factors such as tool geometry, choice of welding parameters, postweld heat treatment (PWHT), workpiece thickness, thermal cycle during welding, and auxiliary cooling also greatly influence the resultant microstructure and strength of the weldment. Influence of above stated factors on FSW characteristics of Al–Li alloys will be discussed in the subsequent sections.

5.2 FSW OF Al–Mg–Li ALLOYS

Al–Mg–Li alloys have moderate strength and are ultralight (density $\approx 2.5 \text{ g/cm}^3$) as compared to Al–Cu–Li alloys. Addition of Mg and Li reduces the density of aluminum alloys. Al–Mg–Li alloys were mainly developed in Russia [2,3]. Thermally stable alloy 1424 is a heat treatable Al–Mg–Li–Zr alloy that was developed out of 1420 and

Figure 5.1 (A) Dark field TEM image showing high density of δ′ precipitates in Al—Mg—Li alloy 1424 and (B) SADP of area shown in (A) [4]. Source: Reprinted with permission from Elsevier.

1421 alloys [3]. It achieves strength primarily from precipitation of metastable δ' (Al_3Li) phase and solid solution strengthening offered by Mg [2—4]. The spherical δ' precipitate has $L1_2$-ordered structure and is coherent with aluminum matrix (Fig. 5.1) [5,6]. The δ' particles are spherical and have very small interfacial energy of $\approx 10 \text{ mJ/m}^2$ [7]. The 1424 alloy was developed in Russia with an aim to replace heavier 2024 aluminum alloy (density $= 2.78 \text{ g/cm}^3$) in fuselage applications in airplanes [3,8,9]. Sidhar et al. [4] studied the joining of 1424 alloy using FSW. They evaluated the effect of PWHT on microstructural and mechanical property evolution in welds of 1424 alloy [4]. 1424 is a precipitation-strengthened alloy; thus, external cooling mediums can be used to produce welds with different thermal experience during welding. Therefore, Sidhar et al. [4] made two welds: weld 1 (W1) was made on a regular welding bed (steel backing) and ambient cooling as the heat sink, and weld 2 (W2), a water-drenched paint roller traversing behind the welding tool and a copper backing plate, was used to facilitate higher cooling rate and lower peak temperature as compared to weld 1. Further experimental details are explained elsewhere [4].

5.2.1 Hardness and Microstructural Evolution

As mentioned earlier, Al—Mg—Li alloys are strengthened by Al_3Li precipitates. Al_3Li precipitates have low thermal stability as compared to other precipitates (T_1, S, θ') of 2XXX alloys. Fig. 5.2 shows the hardness maps of both the welds in AW and PWHT conditions. The three characteristic zones: WN, TMAZ, and HAZ were observed in

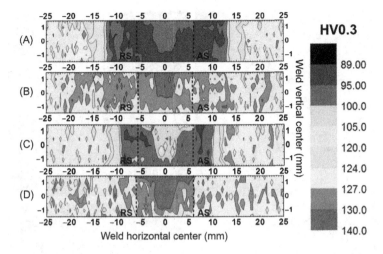

Figure 5.2 Microhardness map of weld cross-section of (A) weld 1 in as-welded condition, (B) weld 1 in PWHT condition, (C) weld 2 in as-welded condition, and (D) weld 2 in PWHT condition [4]. Source: Reprinted with permission from Elsevier.

both the welds. Hardness of both welds in AW condition showed the presence of HAZ, where hardness dropped to 90–100 HV from a base material hardness value of 130 HV [4]. The hardness reduction in HAZ is due to the dissolution of Al_3Li precipitates as the HAZ experiences temperature in range of 200–300°C, which is high enough for dissolution of low thermal stability δ' precipitates. However, the extent of dissolution in HAZ of W2 was lower due to external cooling used in W2. Hardness in the WN in AW condition for both the welds was around 95–105 HV [4]. FSW results in dissolution of precipitates in WN and resulted in decreased hardness as compared to the base material.

After PWHT, hardness in HAZ of both W1 and W2 showed a positive response. Hardness in HAZ improved almost to the level of the base material for both welds. Full recovery of hardness in HAZ during PWHT indicates that the majority of HAZ region did not experience any significant coarsening, and dissolution of precipitates was responsible for decrease in hardness in the AW condition. In WN, hardness of both the welds increased during PWHT and was higher than the base material by 7–11 HV [4]. Hardness in the WN of W2 after PWHT was slightly higher than W1. Sidhar et al. [4] attributed the external cooling used in W2 for higher hardness increase in WN of W2. Fig. 5.3 shows the TEM images of microstructure of WN of W1 and

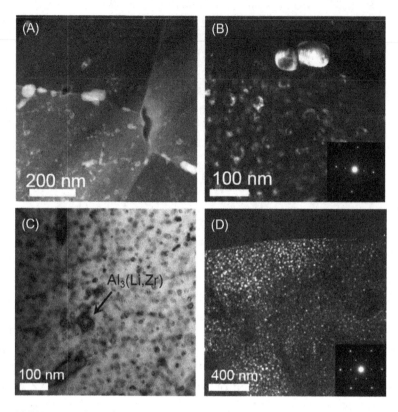

Figure 5.3 TEM images of WN of (A and B) W1 in PWHT condition and (C and D) W2 in PWHT condition [4].
Source: Reprinted with permission from Elsevier.

W2 in PWHT condition. Clearly, a high density of δ' precipitates can be observed in both W1 and W2 samples. Overall, precipitate size in WN of W1 sample was larger compared to WN of W2 and BM microstructure. Grain boundary precipitation was also observed in WN of W1, as shown in Fig. 5.3A. Whereas very few grain boundary precipitates were observed in WN of W2. Extensive grain boundary precipitation in WN of W1 was mainly due to the low cooling rates associated with W1. Sidhar et al. [4], with a combined analysis of DSC and TEM results, concluded that low undercooling and larger residence time at higher temperature associated with W1 led to solute migration to grain boundaries, which favored grain boundary precipitation, whereas external cooling provided in W2 led to higher solute availability for solid solution strengthening and dense precipitation of fine δ' precipitates during PWHT.

In the case of FSW of Al—Mg—Li alloys, critical results such as the positive response of HAZ to PWHT and high strength recovery in WN are contrary to typical cases of FSW of precipitation-strengthened aluminum alloys [4,10]. Typically, FSW of precipitation-strengthened aluminum alloys (2XXX and 7XXX series) results in the knockdown of hardness in HAZ and WN as discussed in earlier chapters. Temperature rise in HAZ during FSW can range in 200—350°C. In FSW of 2XXX and 7XXX alloys, loss of strength in HAZ occurs due to coarsening or dissolution of precipitates occurring at high temperatures. Also, 2XXX and 7XXX alloys possess complex multiprecipitate microstructure, and coarsening occurs very rapidly in HAZ. Even welds made at high speeds experience some level of property degradation in HAZ which, subsequently, shows a negative response to PWHT due to further coarsening of precipitates. However, in the case of Al—Mg—Li alloys, temperatures reached in HAZ are enough to dissolve δ' precipitates. Al—Mg—Li alloys mainly consist of δ' precipitates as the strengthening phase. Also, δ' precipitate has very low interfacial energy and precipitates easily and homogenously throughout the Al matrix. Thus, δ' reprecipitates readily in HAZ and WN upon exposure to PWHT. This explains the positive response of welds of Al—Mg—Li alloys to PWHT.

5.3 FSW OF Al—Cu—Li ALLOYS

FSW of Al—Cu—Li alloys also shows a few characteristics similar to other 2XXX aluminum alloys discussed in earlier chapters. Alloy chemistry and initial microstructure controls the final microstructure and strength of the weld. T_1 (Al$_2$CuLi) is the predominant strengthening phase in Al—Cu—Li alloys. T_1 phase precipitates as plates on {1 1 1} planes in the Al matrix and can be identified from the reflections (reflection at $1/3 < 2\ 2\ 0 >$ and streaks along $< 1\ 1\ 1 >$ direction in Fig. 5.4B) in SADP. L1$_2$ type δ' (Al$_3$Li) precipitate is also observed in Al—Cu—Li alloys. Superlattice reflection in SADP labeled in Fig. 5.4B shows the presence of δ' precipitate. The volume fraction of δ' precipitate is usually low if the Li content is below 1 wt%. θ' and S phases from Al—Cu and Al—Cu—Mg are also observed in Al—Cu—Li alloys, however, the volume fraction of these precipitates is very low. Al—Cu—Li alloys present an extremely complex precipitate system. FSW is a complex thermomechanical joining process. Consequently, joining of Al—Cu—Li alloys using FSW results in complex microstructural evolution in weld cross-section and is discussed in detail in the next section.

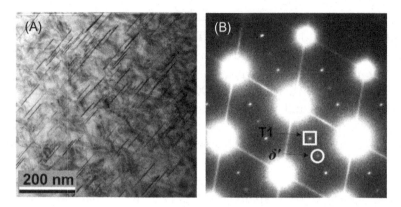

Figure 5.4 (A) TEM image showing T$_1$ precipitates observed in Al−Cu−Li alloys and (B) SADP of area shown in (A) showing reflections of T$_1$ and δ′ precipitates [10]. Source: Reprinted with permission from Elsevier.

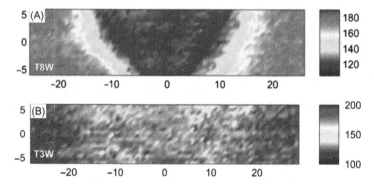

Figure 5.5 Hardness map of weld cross-section of an Al−Cu−Li alloy in (A) T8 temper [12] and (B) T3 temper as starting material [13]. Source: (A) Reprinted with permission from Taylor & Francis. (B) Reprinted with permission from Elsevier.

5.3.1 Microstructure and Hardness Evolution

Microstructural evolution in various zones of welds of Al−Cu−Li alloys is greatly influenced by the choice of initial temper of the material. Fig. 5.5 shows the hardness maps of weld cross-sections of Al−Cu−Li alloy in peak aged (Fig. 5.5A) and underaged (Fig. 5.5B) tempers. The welds cross-section shown in Fig. 5.5 is 10 mm thick. Although, the thickness of material can significantly impact the final strength of the material, basic physical metallurgy can explain the microstructural evolution in weld of Al−Cu−Li alloy in industrially utilized thicknesses. In this section, microstructural evolution in various zones of Al−Cu−Li alloys is discussed in conjunction with hardness across the weld cross-section.

The thermal profile and plastic deformation in the weld zones during FSW change the microstructure of the material and hence affect its mechanical properties. Commonly, a hardness drop of $\sim 60-80$ HV in the TMAZ and SZ of welds is observed. In Al–Cu–Li alloy system, the strength increase due to grain size reduction during FSW and large increase in strength are achieved during PWHT. The change in precipitate density and morphology observed in WN, HAZ, and TMAZ causes a drastic change in microstructure and properties of the weld. In case of welds of peak aged material, the hardness shows a "W" or "U" shaped profile. In case of welds of thick sections, the hardness in the weld center is slightly higher as compared to the HAZ which result in W-shaped hardness curve. The decrease in hardness in HAZ is usually higher for thicker welds resulting in a prominent W shape. Whereas, in case of welds of thin sheets, the hardness profile shows a U shape. Therefore, in AW condition, the weakest zone in case of welds of peak aged Al–Cu–Li alloys varies from WN to HAZ depending upon the thickness of material and welding parameters. This characteristic is different from the case of peak aged Al–Cu–Mg alloys, where HAZ is consistently recognized as the weakest zone. Pouget and Reynolds [11] investigated the microstructure of FSWed 2050 alloy. They studied the hardness change during PWHT and observed a considerable recovery in WN and HAZ. The shape of hardness changed from U to W after PWHT.

In case of welds of underaged Al–Cu–Li alloys, the hardness across the weld cross-section shows a small variation close to BM hardness in underaged condition as evident from Fig. 5.5B. This characteristic is strikingly different as compared to the case of underaged Al–Cu–Mg alloys. In underaged Al–Cu–Mg alloys, two distinct low hardness zones are observed. The kinetics of T_1 precipitates in Al–Cu–Li system are slower than S precipitate in Al–Cu–Mg alloy system. Thus, extensive coarsening of T_1 precipitates in welds of underaged Al–Cu–Li alloys does not occur, whereas S precipitate is known to coarsen rapidly in HAZ of underaged Al–Cu–Mg alloys.

5.3.1.1 Weld Nugget
Similar to the case of other 2XXX aluminum alloys, temperature in WN during FSW of Al–Cu–Li alloys can range in $400-550°C$ as experimentally and numerically demonstrated by researchers [10,14]. As a result of intense plastic deformation at high temperatures, WN undergoes

recrystallization. WN is characterized by equiaxed, recrystallized grains with size 1−15 μm depending upon welding parameters. Furthermore, such high temperatures are enough to dissolve the strengthening precipitates in WN. Shukla [14] studied the microstructural evolution in FSWed 2195-T8 alloy in thin gage (1 mm). In TEM analysis, they found that SADP taken from different grains did not reveal the presence of the dominant T_1 precipitates or θ' and S precipitates which form in very low volume fraction. The TEM images shown in Fig. 5.6 clearly show that the WN experiences dissolution of precipitates during FSW. Similar attribute of WN of various Al−Cu−Li alloys in AW condition was reported in other studies [10,12]. Shukla [14] also demonstrated that in the case of thin sheets, the process parameters have weak effect on the microstructure evolved in WN during FSW and the hardness remains similar. Shukla [14] used a set of wide range of welding parameters (1800 RPM−300 mmPM, 2000 RPM−225 mmPM, 2200 RPM−150 mmPM, and 2400 RPM−75 mmPM) for 1-mm-thick sheets of 2195-T8 alloy. In all of these welds, the microstructure was similar to the images shown in Fig. 5.6. Shukla [14] also reported a few coarse precipitates with a structure similar to T_1 phase, which can form during cooling of FSW process. δ' (Al$_3$Li) precipitates can also form in WN during natural aging due to low activation and interfacial energies. The initial temper of the material (T3 or T8) does not have a significant influence on the microstructure in WN as temperature in WN is high enough to create super-saturated solid solution. Geuser et al. [12] also studied the precipitate distribution and microstructure of a FSWed 2050-T8 Al−Li alloy using SAXS, TEM, and hardness testing. They observed complete dissolution of T_1 precipitate in nugget after FSW which occurred due to

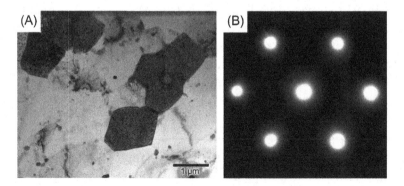

Figure 5.6 (A) TEM image showing the microstructure of WN of Al−Cu−Li alloy in AW condition and (B) SADP of area shown in (A) showing absence of strengthening precipitates [14]. Source: Reprinted with permission from Dr. A.K. Shukla.

the fact that typically welding temperature in FSW of aluminum alloys lies in the range of 400–500°C, well above dissolution temperature of the precipitate. Also, clusters or GP zones were observed in the nugget after welding where complete dissolution of T_1 had occurred. Overall, similar to the case of other 2XXX aluminum alloys, WN of Al—Cu—Li alloys in the AW condition shows a low hardness region owing to the complete dissolution of strengthening phases.

5.3.1.2 Thermomechanically Affected Zone

As explained in earlier chapters, TMAZ is a fine transition zone between WN and HAZ, consisting of highly deformed grains. It is characterized by dense dislocation structures with density usually higher than that of base material. In AW condition, this zone also experiences complete dissolution of strengthening precipitates similar to WN as the temperature in TMAZ remains close to WN temperature. However, TMAZ in Al—Cu—Li alloys shows a different character as compared to other 2XXX aluminum alloys. T_1 precipitate is known to have enhanced kinetics in the presence of dislocations [10,15]. In case of Al—Cu—Li alloys, enhanced precipitation of T_1 precipitates during PWHT in TMAZ is observed due to high dislocation density. This results in higher hardness recovery in TMAZ after PWHT as clearly evident in TMAZ zones in all the hardness plots shown in a later section in Fig. 5.8.

5.3.1.3 Heat Affected Zone

HAZ is usually the weakest region in welds of Al—Cu—Li alloys in AW condition. HAZ is critical when the initial temper of the material is peak aged (T8). The lowest hardness in AW condition forms just outside of TMAZ, near TMAZ—HAZ boundary. Microstructural variation across the HAZ follows the gradient of temperature across the HAZ. The region of HAZ closer to TMAZ, where temperature is usually above 300°C, results in partial dissolution of strengthening phases as shown in Fig. 5.7A–C. Also, having experienced relatively high temperature, coarse T_1 and $L1_2$-type precipitates can also form in this region as shown in Fig. 5.7C. SADP images shown in Fig. 5.7A and B also confirm the presence of T_1 and $L1_2$-type precipitates. Shukla [14] observed coarse plates of T_1 as thick as 10 nm and as large as 140 nm in this region (Fig. 5.7C) as compared to average length of 86 nm in BM in T8 condition [14]. Whereas, the region of HAZ further away from TMAZ experiences lower temperatures in the range of 200–250°C. Such temperatures do not result in dissolution of T_1 precipitates, but significant level of

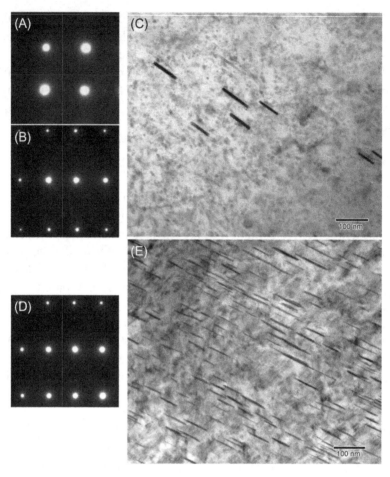

Figure 5.7 (A and B) SADP patterns, (C) BF TEM image of a region in HAZ close to TMAZ of an Al—Cu—Li alloy in AW condition, (D) SADP pattern, and (E) BF TEM image of a region in HAZ further away from TMAZ and closer to BM of an Al—Cu—Li alloy in AW condition [14]. Source: Reprinted with permission from Dr. A.K. Shukla.

coarsening of T_1 phases can occur. The extent of coarsening in this region highly depends on the exposure time at high temperatures which is usually controlled by the welding speed and cooling rates. Fig. 5.7D and E shows the microstructure of a similar region studied by Shukla [14]. It shows high density of T_1 precipitates in BF image (Fig. 5.7E). SADP shown in Fig. 5.7D also indicates the presence of T_1 and $L1_2$ precipitates (β' and δ' precipitates). Shukla [14] quantitatively showed that average length of T_1 precipitates increased in this region as compared to BM precipitates, which signifies the limited coarsening in this region. It is important to note that the study presented by Shukla [14] was conducted on 1-mm-thick sheets. In such low thickness material, welding speed as high as 300 mmPM can be achieved which result in very high cooling rates. However, in case of welding of thicker sections (>6 mm), optimized welding speeds decrease with increase in thickness of the workpiece, and thus the cooling rates reduce. Therefore, as discussed earlier, the extent of coarsening of T_1 precipitates in HAZ generally increases with increase in thickness of the material. This results in increase in width of HAZ in welding of thicker sections.

As discussed in the case of FSW of Al–Cu–Mg alloys (alloy 2024), the HAZ shows the presence of coarsened S precipitates. This characteristic was observed in FSW of 2024 in the T3 condition. The kinetics of S precipitates are high enough for precipitation to occur in short thermal exposure time during FSW. In the case of Al–Cu–Li alloys, FSW of underaged (T3 or T4) alloy does not result in extensive precipitation and coarsening in HAZ. Malard et al. [13] demonstrated that the HAZ of underaged alloy in AW condition contains only GP zones which form during postweld natural aging. HAZ of underaged Al–Cu–Li alloy experiences only a reversion cycle. Therefore, HAZ of underaged Al–Cu–Li alloy has low significance as Al–Cu–Li alloys are intended for high strength applications and thus PWHT is carried out.

5.3.2 Effect of Alloy Chemistry and Postweld Heat Treatment

Alloy chemistry plays an important role in determining the type, volume fraction, and size of precipitates in an alloy system. Within Al–Cu–Li alloys, Cu–Li ratio determines the relative volume fraction of T_1 and δ' precipitates. The role of Cu–Li ratio has already been discussed in earlier chapters. Overall, amount of alloying elements is controlled to design property-specific alloys. Third generation of Al–Li alloys was developed with reduced Li concentration (<1.8 wt%) to overcome the shortcomings of second generation of Al–Li alloys. Similarly, 2199 alloy was developed with improved combination of mechanical properties and corrosion resistance as compared to its predecessors [1]. 2199 alloy contains lower Cu and higher Li and Zn as compared to other high strength third-generation Al–Li alloys (2195, 2050, etc.). Absence of Ag in alloy 2199 is another major difference as compared to alloys 2195 and 2050. Ag stimulates nucleation and growth of T_1 phase [16,17]. Due to its different chemistry, 2199 alloy is strengthened by both T_1 and δ' precipitates, whereas alloys 2195 and 2050 are strengthened by only T_1. Higher Li content results in precipitation of high volume fraction δ' phases.

Sidhar and Mishra [10] conducted a comparative study on aging kinetics in welds of alloys 2199 and 2195 during PWHT using various experimental techniques such as hardness measurements, DSC, and TEM. Samples from welds of both the alloys were heat treated for various number of hours leading up to 100 hours at 160°C to study the aging kinetics of various precipitates in various zones. The hardness profiles for all the samples from welds of alloys 2195 and 2199 are shown in Fig. 5.8. Overall, it is clear from the hardness plots that the aging kinetics in WN and HAZ of Ag-free 2199 alloy were sluggish as compared to that of Ag-containing 2195 alloy. In WN in the AW condition, both the alloys showed solutionized microstructure in TEM examination. WN of 2195 alloy, in which T_1 precipitate is the predominant strengthening phase, showed limited strength improvement after aging at 160°C for 16 hours. Whereas, WN of 2199 showed nearly no response to first aging treatment at 160°C for 16 hours. Sidhar et al. [10] concluded that the absence of Ag and low dislocation density in WN, both of which promote nucleation of T_1 precipitates, in 2199 alloy were responsible for extremely slow aging response. Whereas, in the case of 2195 alloy, although the dislocation density in WN was low, presence of Ag resulted in enhanced aging response as compared to 2199 alloy.

Another significant effect of alloy chemistry can be noted by comparing the HAZ of both the alloys. In TEM analysis, HAZ of both the alloys also showed partial dissolution of T_1 precipitates. However, 2199 alloy, which contains a significant volume fraction of low thermal stability δ' precipitates, showed a nearly full dissolution of δ' phase and lesser

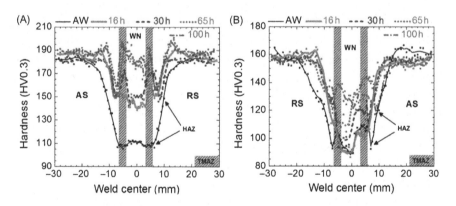

Figure 5.8 Hardness plot of welds of (A) 2195 alloy and (B) 2199 alloy subjected to various PWHT [10].
Source: Reprinted with permission from Elsevier.

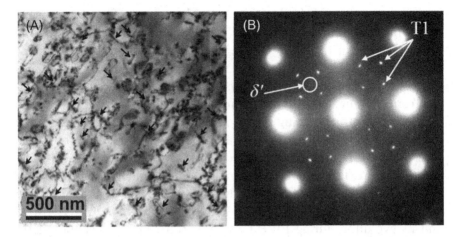

Figure 5.9 TEM image of (A) HAZ of weld of alloy 2199 in AW condition and (B) SADP taken along [1 0 0] zone axis of area shown in (A). T_1 precipitates are marked with arrows in (A). Source: Reprinted with permission from Elsevier.

dissolution of T_1 phase. The BF and SADP images shown in Fig. 5.9B clearly indicate that amount of δ' precipitates remaining in HAZ was very low. Sidhar and Mishra [10] concluded that the absence of Ag and low dislocation density in WN of 2199 alloy resulted in extremely sluggish aging kinetics during PWHT. Whereas, even though dislocation density was quite low in WN of 2195 alloy, Ag acted as nucleation agent and resulted in far better aging kinetics in 2195 alloy. Another study involving detailed microstructural analysis of 2199 alloy using SAXS was conducted by Steuwer et al. [18]. Their analysis showed that some degree of coarsening and partial dissolution of T_1 precipitates, which were initially present in the alloy, in HAZ and TMAZ occurred. They reported a "U" shaped hardness curve contrary to the typical "W" shaped hardness curve for peak aged Al—Cu—Li alloys. Steuwer et al. [18] also observed the dissolution of δ' precipitates in nugget and coarsening of T_1 precipitate in HAZ owing to the drop in hardness. However, it is clear from the hardness results shown in Fig. 5.8 that W shape hardness can also form in this alloy and the variation in results can be due to the choice of welding parameters in both the studies.

PWHT can further alter the microstructure and mechanical properties of the weld. The response of microstructure in various zones of the weld towards PWHT depends on the initial microstructure of the workpiece. WN undergoes complete dissolution of strengthening phases during FSW and results in similar microstructure irrespective of initial temper, provided other factors are kept same. Therefore, micro-structural evolution during PWHT of WN is insensitive to the initial microstructure of the weld. On the other side, microstructure in HAZ is highly dependent on the initial temper of the material. HAZ in a weld of a peak aged material shows a gradient of microstructure start-ing from high degree of dissolution of precipitates close to TMAZ, to coarsening of strengthening phases away from TMAZ. Whereas, in the case of HAZ of underaged material, the thermal cycle only results in reversion of GP zones thereby essentially producing underaged mate-rial. Therefore, it is important to understand the microstructural evolu-tion in various zones of FSW during PWHT. Malard et al. [13] studied the microstructural distribution in the cross-section of a 2050-T34 weld and its evolution during PWHT. They performed a systematic micro-structural mapping by using a combination of SAXS, microhardness, TEM, and DSC. Fig. 5.10A shows the microhardness measurement across the weld in various PWHT conditions. Sample named as

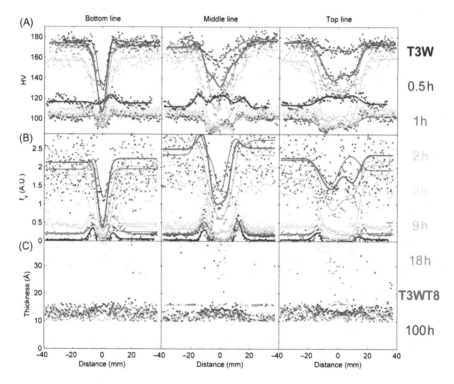

Figure 5.10 (A) Microhardness and (B) and (C) SAXS measurement taken at three horizontal planes of a weld cross-section of alloy 2050-T3 subject to PWHT for various number of hours at 155°C [13]. Source: Reprinted with permission from Elsevier.

"T3WT8" was subjected to a peak aged heat treatment of 30 hours at 155°C as identified by Malard et al. [13]. Similarly, other legends in Fig. 5.10 show the number of hours of heat treatment at 155°C for the sample under consideration. Fig. 5.10B and C shows the volume fraction and thickness of precipitates, captured using SAXS experiments, across the weld cross-section of underaged 2050 alloy aged to various number of hours at 155°C.

Al–Cu–Li alloys in T3 condition contains only GP zones. Malard et al. [13] showed that the microstructure of 2050-T34 alloy after welding consists of a complex spatial distribution of solute clusters (GP zones). However, these variations result in small changes in hardness across the weld cross-section as shown in Fig. 5.5B. Malard et al. [13] observed the presence of a small amount of plate-type precipitates in TMAZ in AW condition, which resulted in slightly higher hardness in TMAZ. High density of dislocations in TMAZ and high temperature experienced during FSW resulted in the precipitation of plate-type phases in such a short time. Overall, weld cross-section of underaged alloy in AW condition results in small variation in hardness profile which can flatten out after a few weeks of natural aging.

After PWHT, different zones of the weld exhibit different aging response. As clearly evident from hardness profiles shown in Fig. 5.10A, BM showed peak hardness after 30 hours of aging (labeled as T3WT8). Hardness increase in WN was slowest across the weld cross-section. Even after aging for 100 hours, WN could not show hardness similar to BM which showed peak hardness in 30 hours of aging. Malard et al. [13] ascertained the lower dislocation density in the WN as the key reason for slower aging kinetics in WN. Absence of dislocations (or low dislocation density) reduces the nucleation kinetics of T_1 precipitates which are known to be favored by the presence of dislocations. Malard et al. [13], with a combined analysis of DSC, SAXS, and TEM, demonstrated that during long aging hours, slow precipitation of θ' (Al_2Cu) precipitates is favored in WN as compared to early precipitation of T_1. Malard et al. [13] experimentally clarified the hypothesis of low dislocations being the key reason for low hardness in WN after PWHT. They applied 10% plastic strain to a sample in AW condition, followed by peak aging treatment of 30 hours at 155°C [13]. This sample showed a homogenous distribution of hardness (Fig. 5.11) across the weld cross-section. At the top of the weld, the hardness was observed to

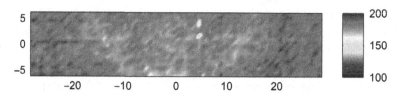

Figure 5.11 Microhardness map of weld cross-section of weld of 2050-T3 alloy subjected to 10% plastic strain followed by PWHT at 155°C for 30 hours [13]. Source: Reprinted with permission from Elsevier.

be completely homogeneous and recovered to the level of BM hardness. Whereas, at the middle and lower sections of the WN, a moderate hardness improvement but lower than BM hardness was observed, and the bottom of the weld was observed to show most pronounced difference as compared to BM (Fig. 5.11). The lower recovery of hardness towards the bottom of the weld was due to the loss of solute to the coarse insoluble intermetallic particles and grain boundaries. In the case of welding of thick sections, the cooling rates and peak temperature reduce along the depth which provides enough time and temperature for solutes migration from solid solution to intermetallic particles. Malard et al. [13] experimentally obtained the area fraction of coarse intermetallic particle at various locations in the weld cross-sections. It was observed that fraction intermetallic particles were much higher towards the bottom of the weld as compared to the top of the weld which showed intermetallic content similar to that of BM. Top of the weld has highest cooling rates and thus the solute loss is minimized, resulting in better hardness recovery during PWHT as compared to the bottom section of the weld as shown in Fig. 5.11.

REFERENCES

[1] R.J. Rioja, J. Liu, The evolution of Al—Li based products for aerospace and space applications, Metal. Mater. Trans. A 43 (2012) 3325—3337.

[2] I. Fridlyander, Aluminum alloys with lithium and magnesium, Metal Sci. Heat Treat. 45 (2003) 344—347.

[3] I. Fridlyander, L. Khokhlatova, N. Kolobnev, K. Rendiks, G. Tempus, Thermally stable aluminum—lithium alloy 1424 for application in welded fuselage, Metal Sci. Heat Treat. 44 (2002) 3—8.

[4] H. Sidhar, N.Y. Martinez, R.S. Mishra, J. Silvanus, Friction stir welding of Al—Mg—Li 1424 alloy, Mater. Des 106 (2016) 146—152.

[5] J.W. Martin, Aluminum—lithium alloys, Ann. Rev. Mater. Sci. 18 (1988) 101—119.

[6] S.C. Jha, Precipitation processes in aluminum—lithium—magnesium alloys, ProQuest Dissertations and Theses (1987).

[7] A. Deschamps, C. Sigli, T. Mourey, F. De Geuser, W. Lefebvre, B. Davo, Experimental and modelling assessment of precipitation kinetics in an Al—Li—Mg alloy, Acta Mater 60 (2012) 1917—1928.

[8] I. Fridlyander, A. Dobromyslov, E. Tkachenko, O. Senatorova, Advanced high-strength aluminum-base materials, Metal Sci. Heat Treat. 47 (2005) 269—275.

[9] I. Fridlyander, V. Sister, O. Grushko, V. Berstenev, L. Sheveleva, L. Ivanova, Aluminum alloys: promising materials in the automotive industry, Metal Sci. Heat Treat 44 (2002) 365—370.

[10] H. Sidhar, R.S. Mishra, Aging kinetics of friction stir welded Al—Cu—Li—Mg—Ag and Al—Cu—Li—Mg alloys, Mater. Des 110 (2016) 60—71.

[11] G. Pouget, A.P. Reynolds, Residual stress and microstructure effects on fatigue crack growth in AA2050 friction stir welds, Int. J. Fatigue. 30 (2008) 463—472.

[12] F. De Geuser, B. Malard, A. Deschamps, Microstructure mapping of a friction stir welded AA2050 Al—Li—Cu in the T8 state, Philos. Mag 94 (2014) 1451—1462.

[13] B. Malard, F. De Geuser, A. Deschamps, Microstructure distribution in an AA2050 T34 friction stir weld and its evolution during post-welding heat treatment, Acta Mater. 101 (2015) 90—100.

[14] A.K. Shukla, Friction stir welding of thin-sheet, age-hardenable aluminum alloys: a study of process/structure/property relationships, ProQuest Dissertations and Theses (2007).

[15] W. Cassada, G. Shiflet, E. Starke, The effect of plastic deformation on Al_2CuLi (T 1) precipitation, Metal. Trans. A 22 (1991) 299—306.

[16] B.- Huang, Z.- Zheng, Independent and combined roles of trace Mg and Ag additions in properties precipitation process and precipitation kinetics of Al—Cu—Li—(Mg)—(Ag)—Zr—Ti alloys, Acta Mater 46 (1998) 4381—4393.

[17] M. Murayama, K. Hono, Role of Ag and Mg on precipitation of T 1 phase in an Al—Cu—Li—Mg—Ag alloy, Scr. Mater. 44 (2001) 701—706.

[18] A. Steuwer, M. Dumont, J. Altenkirch, S. Birosca, A. Deschamps, P.B. Prangnell, et al., A combined approach to microstructure mapping of an Al—Li AA2199 friction stir weld, Acta Mater. 59 (2011) 3002—3011.

CHAPTER 6

Physical Metallurgy-Based Guidelines for Obtaining High Joint Efficiency

Aluminum 2XXX series encompass the most complex precipitation-strengthened alloys. In earlier chapters, we discussed the microstructural and hardness evolution in various 2XXX alloys in different tempers and postweld heat treatment (PWHT) conditions. Overall, the performance of the joints of 2XXX alloys is influenced by the initial microstructure and welding parameters. Both physical metallurgy- and design-based approaches can be utilized to improve the overall performance of the weld of 2XXX alloys. The WN, TMAZ, and HAZ are in different microstructural states based on the process parameters and thermal boundary conditions. In this chapter, we provide a short discussion to outline some guidelines as well as highlight some potential future approaches to achieve high joint strength.

As discussed in earlier chapters, FSW of thicker material produces lower joint strength as compared to thin sheets. Welding speed decreases with increase in thickness of material and results in higher level of degradation in the HAZ. Best properties in the WN can be achieved by attaining near solutionization temperature accompanied by high cooling rates to retain the solutes in solid solution. However, large thermal gradient along the depth in the WN of thick sections results in large difference in peak temperatures at the top and bottom of the WN and results in heterogeneous properties. In such cases, higher deterioration in properties is observed at the bottom part of the weld. Whereas, low peak temperature and high cooling rate are needed in the HAZ for optimum results. Use of a high thermal conductivity backing plate (e.g., Cu, Al, etc.) can significantly improve the cooling rates. However, this will result in higher thermal gradient along the depth in the WN. Use of a low thermal conductivity material (e.g., ceramics) as backing can restrict the loss of heat towards backing and reduce the thermal gradient along the depth in weld cross-section.

Friction Stir Welding of 2XXX Aluminum Alloys Including Al–Li Alloys.
DOI: http://dx.doi.org/10.1016/B978-0-12-805368-3.00006-6

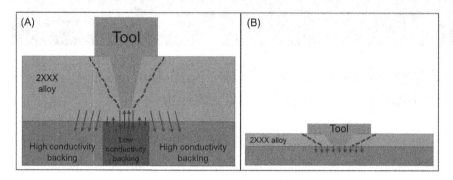

Figure 6.1 Schematic showing arrangement of backing plate suggested for FSW of (A) thick sections and (B) thin sheets of 2XXX alloys.

However, the use of low thermal conductivity backing material will lead to increase in the overall heat content in the HAZ, which will result in higher loss of strength in the HAZ. Thus, a composite backing plate arrangement consisting of high thermal conductivity backing for HAZ and low thermal conductivity material backing for WN will result in optimum properties across the weld cross-section. A schematic explaining the arrangement of backing plate in FSW of thick section is shown in Fig. 6.1A.

In case of thin sheets, thermal gradient is negligible as the entire WN lies in close proximity of tool shoulder which produces the major fraction of heat during FSW. Therefore, a single high thermal conductivity backing can be used to achieve higher cooling rates and to reduce the thermal imprint in the HAZ to obtain enhanced properties across the weld cross-section. Fig. 6.1B shows a schematic showing the backing plate arrangement used during FSW of thin sheets of 2XXX alloys. Additionally, auxiliary cooling systems, such as liquid cooled backing plate targeted at the HAZ, air or spray cooling behind the tool aimed at the WN and HAZ, can further expand the joint strength window of FSW of 2XXX alloys. Similarly, intense heating sources such as laser can be used to presoften the material in front of the welding tool to achieve higher welding speeds. Laser preheating also decouples the heat input requirements and the material flow. This can open up the process window and at this stage not much has been reported on hybrid laser–FSW approach. Note that laser heating will produce its own thermal gradient along the depth of material further complicating the process. Also, a general concern with laser heating of aluminum alloys is the reflectivity and associated energy loss, which in turn

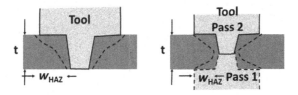

Figure 6.2 An illustration of single pass approach versus double pass approach for thick plates. Note the change in the overall HAZ path in the double pass weld. The associated fracture path during transverse loading will change as the crack path becomes more complicated for the double pass weld [1]. Source: Reprinted with permission from Elsevier.

decreases the process efficiency. The advantages of having a larger process window and coupled joint efficiency enhancement need to be weighed against the economic and other manufacturing concerns.

From mechanistic approach, the angle of HAZ plane (around 45°) with respect to loading axis during tensile test often favors and promotes fracture path. This impacts the performance of welds. A simple estimate of the angle of fracture path can be estimated as $\tan \frac{t}{W_{HAZ}}$ as shown in Fig. 6.2 [1]. This aspect can be appreciated further by consideration of FSW of thick plates by single pass or double pass approach. Fig. 6.2 shows the change in the overall HAZ pattern after the double pass weld. Note that the traverse rate can be higher for double pass weld because of the smaller tool. That again lowers the width of the HAZ region and reduces the extent of deterioration of strength in HAZ. The combined effect would eventually produce higher performance in double pass weld. Of course, from the design perspective the accessibility of the plate for double-side weld is required for such approach. It is important to note that this is not similar to the double-side bobbin tool approach. The use of bobbin tool eliminates the backing plate that extracts heat more efficiently. The impact of these approaches on the distribution of precipitates and PWHT response is quite significant.

REFERENCE

[1] R.S. Mishra, M. Komarasamy, Friction Stir Welding of High Strength 7XXX Aluminum Alloys, Butterworth-Heinemann, UK, 2016. p. 102. ISBN: 978-0-12-809465-5.

CHAPTER 7

Summary and Future Outlook

The joining of high-strength aluminum alloys via solid-state friction stir welding (FSW) has been exceedingly successful. Some of the early adoption of this technology was for aerospace applications by Boeing and NASA. A great extent of knowledge exists on the role of various FSW process parameters in obtaining defect-free high quality welds on the formation of defects and banded heterogeneous microstructure, and finally in creating the process map to guide the future endeavors.

Temperature variation across different weld zones is inherent to the welding techniques, including the relatively low peak temperature in solid-state friction stir welds. As a result, there is variation in both the microstructural features and the corresponding mechanical properties. Therefore, an appropriate control of the thermal fields during and/or after the welding with the help of intrinsic process parameters and/or external heating or the cooling media greatly modifies the results. The thermal fields have to be applied based on good understanding of microstructural evolution in various zones and also the expected nature of evolution. Particular emphasis was given to the difference in precipitation sequence in 2XXX alloys compared to 7XXX series alloys (subject of the sixth volume in this short book series). For example, the low hardness regions in 2XXX alloys differ within the Al–Cu alloys and also from other alloy series like 7XXX and 6XXX. Implication of this can be gaged from an early use of Al 2219 alloy by NASA that has $\sim 6\%$ Cu. This alloy has poor joint efficiency as compared to some of the recent Al–Li alloys. Therefore, care should be taken in selection of the alloy chemistry as well as the postweld heat treatment procedures. It should be based on the physical metallurgy applicable to the particular zone of interest. The most critical aspect is the development of new heat treatment procedures by considering the microstructural heterogeneity and variation, particularly in order to achieve a high joint efficiency structure.

Friction Stir Welding of 2XXX Aluminum Alloys Including Al–Li Alloys.
DOI: http://dx.doi.org/10.1016/B978-0-12-805368-3.00007-8

The transverse mechanical properties completely reflect the variation in the microstructure across the weld nugget and the width of various zones. As expected, the plastic strain localization and eventual sample failure occur along the lowest hardness region. Transverse-weld specimen elongation detoriates due to localization which can be avoided by having smaller difference between the minima and maxima, and also by manipulating the width of the lowest hardness region. Research is needed on the mechanics of flow localization and its dependence on the geometrical nature of strength gradient. Some key results have shown that lower temperature aging of the weld leads to higher recovery of the strength across the weld regions and the width of various zones can be manipulated. While strength recovery is attained, any negative impact on the corrosion properties needs to be avoided.

For high-strength Al–Cu–Mg–Li alloys, the use of T3 or T8 base tempers creates additional possibilities. The role of dislocations on precipitation is a very important difference when compared with 7XXX series alloys and other aluminum alloys. Strategies that can retain higher dislocation density in the weld nugget and thermo-mechnical affected zone, coupled with the thermal management of heat affected zone region, have the best outcome possibilities. Significant opportunities exist in this area for further research. Of course, the approach also depends on the thickness of material. The joint efficiencies decrease to quite low values for thicker materials. Obviously, the improvement in the overall structural efficiency towards 100% joint efficiency will be enormous. Based on the current understanding of the FSW process, both the processing parameters and the postwelding treatment conditions have to be systematically investigated to get ∼100% joint efficiency.

Furthermore, the composition of the alloys can be adjusted to introduce microstructural features that are more stable compared to the conventional 2XXX alloys. This can be a long-term approach to fully exploit the attributes of FSW. The current 2XXX alloys include many alloys which are specifically developed for specific applications. For example, alloys that have higher copper content than the solubility of copper at the temperature experienced during FSW will lead to inferior microstructure with coarser precipitates. Effort towards alloys designed for FSW will be groundbreaking and provides significant flexibility to designers.

Printed in the United States
By Bookmasters